true green home

true green home

100 inspirational ideas for creating a green environment at home

Kim McKay and Jenny Bonnin

NATIONAL GEOGRAPHIC

WASHINGTON, D.C.

Published by the National Geographic Society 2009

First published in Australia by ABC Books in 2008
for the Australian Broadcasting Corporation.

Copyright © 2008 True Green (Global) Pty Ltd

ISBN: 978-1-4262-0399-2

Library of Congress Cataloging-in-Publication Data available upon request.

Design, layout, and select images by Marian Kyte

A percentage of proceeds from the sale of *True Green Home*
benefits Clean Up the World.

True Green Home has purchased carbon credits to neutralize
emissions produced by the printing of this book.

True Green ® is a Trademark of True Green (Global) Pty Ltd.

Founded in 1888, the National Geographic Society is one of the largest nonprofit
scientific and educational organizations in the world. It reaches more than 285 million
people worldwide each month through its official journal, NATIONAL GEOGRAPHIC, and its
four other magazines; the National Geographic Channel; television documentaries;
radio programs; films; books; videos and DVDs; maps; and interactive media. National
Geographic has funded more than 8,000 scientific research projects and supports an
education program combating geographic illiteracy.

For more information, please call
1-800-NGS LINE (647-5463)
or write to the following address:

National Geographic Society
1145 17th Street N.W.
Washington, D.C. 20036-4688 U.S.A.

www.nationalgeographic.com

Printed in Italy on recycled paper.

contents

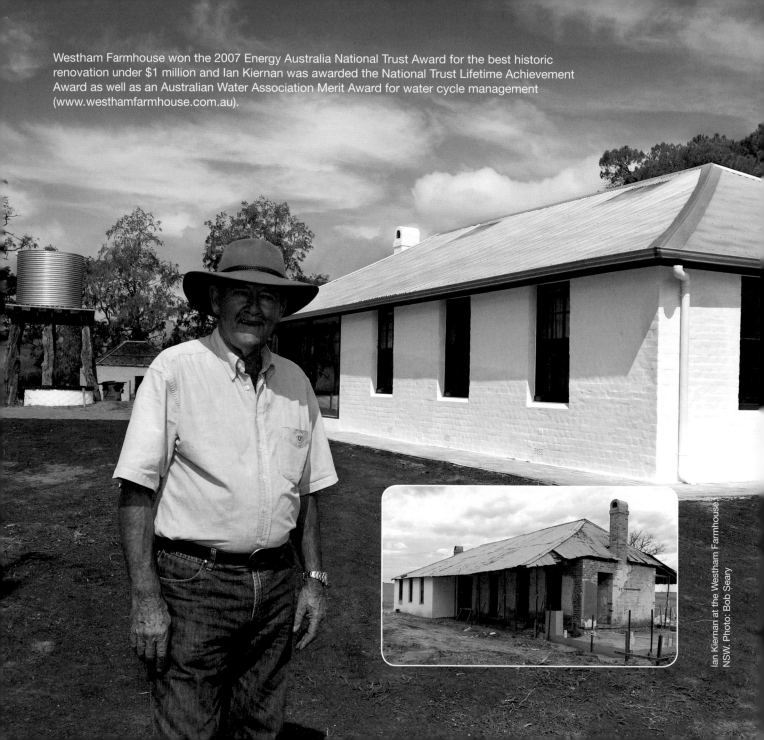

Westham Farmhouse won the 2007 Energy Australia National Trust Award for the best historic renovation under $1 million and Ian Kiernan was awarded the National Trust Lifetime Achievement Award as well as an Australian Water Association Merit Award for water cycle management (www.westhamfarmhouse.com.au).

Ian Kiernan at the Westham Farmhouse, NSW. Photo: Bob Seary

Building, and the satisfaction of creating or restoring, is in my blood. In my youth I built canoes, boats and even furniture and soon embarked on a roller coaster of a career in the building industry.

By the mid '60s I had my own company in Sydney, Australia, building everything from factories to commercial buildings as well as extensions to houses. I'd buy and renovate derelict houses in the then neglected run-down areas of inner Sydney, installing new services and finding new uses—inadvertently preserving much of the city's early working-class heritage.

We had a staggering portfolio of nearly 400 row houses, commercial blocks and restaurants and when the credit squeeze of 1974 struck, I lost the lot! I left the building dream behind, and sailed out of Sydney Harbour for a year's adventure that took me on a life-changing voyage to 36 of the most beautiful Pacific Islands and introduced me to the joys of solo sailing.

In 1986, I competed in the BOC Challenge (an around-the-world yacht race) sailing the 60-foot (18m) *Spirit of Sydney*. One of the competitors, an American, Mark Schrader, encouraged us to hold our plastic waste on board for proper disposal when we reached shore as part of a school environmental project. Through this simple act, I started to see plastic everywhere—and realized just how much waste was ending up in the world's oceans. The builder who had once dumped his old roofing iron in the ocean had seen the light!

Returning home, I saw Sydney Harbour with new, clearer eyes and decided to do something about the state of my own backyard, enlisting the help of Kim McKay, the co-author of this book and co-founder of Clean Up, along with a group of friends.

The rest, as they say, is history. Clean Up Sydney Harbour in 1989 was followed by Clean Up Australia and a few years later, Clean Up the World. Twenty years down the track, the environment and the impact of global warming has become the biggest challenge facing us all.

A fascination with architecture and historic buildings has always stayed with me and a few years ago, a friend showed me Westham Farmhouse, near Bathurst in rural New South Wales. I was awestruck by this example of convict-built Australian pioneer heritage, and even though the house was virtually derelict, I bought it.

Today the farmhouse and outbuildings are restored incorporating all environmental services circa 1830, overlaid with world's best environmental practices circa 2008. There's total water cycle management, solar power and hot water, biological waste management, recycled building materials and native plant regeneration. Westham's restoration has been a labor of love and is an example of how you can integrate environmental products and services into a working building.

True Green Home, created by Kim and fellow Clean Up director Jenny Bonnin, provides easy-to-follow tips on what you can do, whether you are renovating or building from scratch. Be inspired by what you can achieve—it could be the start of the change our world needs.

introduction

Kim McKay and Jenny Bonnin

Since the release of our first book, *True Green: 100 Everyday Ways You Can Contribute to a Healthier Planet*, in 2007, environmental issues have taken center stage globally. It's hard to escape news of the latest scientific findings on climate change, or the most recent way to reduce your carbon emissions, or which car to buy or eco-holiday to take. There's so much advice that it can be confusing.

Some commentators are even claiming that now there is too much information, and that this may contribute to a green backlash; that the "noise" created around these serious issues is in danger of overwhelming the public.

"Green noise" was recently defined by a *New York Times* writer as "the static caused by urgent, sometimes vexing or even contradictory information played at too high a volume for too long." Everyone seems to be saying, "Just tell me what I need to know, just give me the cheat sheet."

We're hoping *True Green Home: 100 Inspirational Ideas for Creating a Green Environment at Home* will be the cheat sheet you're looking for.

Building or renovating a home is one of the biggest financial undertakings you'll face in your life and now more than ever, it's important to get it right. With America having among the world's highest per-capita greenhouse emissions, everyone needs to help reduce our carbon footprint, and there's no better place to start than at home. Households make up almost one-fifth of our greenhouse gases and the simple tips in this

book should help everyone take steps that will lead to a reduction in those emissions.

We've both had experience at renovating homes—most recently a family row house (Jenny's) and a small apartment (Kim's), both near the inner city in Sydney, Australia where we live. We have tried to incorporate many of the practices and principles we've learned during these renovations, and others, into this book.

Jenny's True Green Home Story

I'm a passionate renovator and have always supplemented my income by performing small building and decorative projects on inner-city row houses. The challenge has been to find properties where we can add value. Could parking be added via rear lane access, could a third bedroom fit into the roof cavity, was there any space to reconfigure the kitchen to include an internal laundry, second bathroom and a larger casual living area?

While everything met local council standards and appropriate architectural limitations, the main selling point was that the property presented extremely well, with clean lines and decorative touches from a "I could easily live here and I don't need to do anything!" perspective.

Times are changing: green is definitely the new

black when it comes to marketing homes and smart investments are getting smarter—literally. "Smart houses," "green homes" and "sustainable lifestyle communities" are popping up everywhere.

I took all this on board, along with my passion and determination to make a difference, to be the change I wanted to see, and embarked on working with an architect to create a new green home.

It's very easy to find out about all this from the web resources at the back of the book, but instinctively we all know what to do: we need to decide to live a less extravagant, more simple life, ditching unnecessary lifestyle props like air-conditioning and heated floors when you live in a temperate climate. It's easy to achieve natural airflow and temperature control by working out how to place your house on the land, where the windows are, and through the use of louvers, blinds and fans.

Energy saving is a great conversation starter—it's fashionable to know the solar water heater or photovoltaic system you have chosen, or how much rain is needed this season to accomplish water savings with a newly installed water tank. "Going off the grid" doesn't mean you're dropping out of society but it does mean you are thinking in the right way and developing a sense of self-sufficiency.

I now spend evenings searching suppliers for recycled timber or sustainable timber for the floors, for low- and zero-VOC paint color charts, energy-efficient appliances and native plants for the new garden.

When this house goes on the market, it won't be only the design and decoration that will be the key for potential buyers. A sustainable house is the new trend and soon will become the standard—a "true green home revolution."

Kim's True Green Home Story

I'm all about keeping my lifestyle simple, and with all the traveling I've been doing for work for the past 20-plus years, I decided that apartment living was the most practical for me. A few years back I found an old apartment with good bones and set about making it as energy-friendly as possible.

I have instant gas hot water, which saves on space and energy; energy-saving appliances; windows that remain open to take advantage of cross-breezes in summer; and minimal overhead lighting—using lamps with energy-efficient bulbs to cut down on electricity usage. I used so little gas in the first 18 months that the gas company didn't even send me a bill! My simple renovation was cost-efficient too, as we kept the original floorboards, fixtures and fittings where possible.

I use gas heating during winter and one simple unit heats the entire apartment—I have only four rooms. My next challenge is to install a water tank and maybe even solar panels on the building—I better get in touch with the owner's cooperative to see what can be achieved.

My home is compact and cozy. I call it my "cubbyhouse" and it's my retreat at the end of a long week, somewhere I can rest comfortably, knowing that my carbon footprint is shrinking all the time!

Have fun with *True Green Home* and let us know about your ideas for creating a sustainable home.

www.betruegreen.com

home green home

1

location, location

This golden rule of real estate also applies to creating an eco-friendly home. By considering your property's proximity to public transportation, schools, stores and your work, you can significantly reduce your dependency on your car and add serious value to your home. Leaving the car at home and catching public transportation, riding your bike to work and walking the kids to and from school all contribute to a healthier you and a healthier world. Whether you're renting, renovating or building your home, look for suburbs with established gardens and tree-lined streets that have good access to parks and open spaces to enjoy green vistas and increase the quality of your daily life.

Courtesy Environa Studio. Photo: Tim Wheeler Studios

2

supersize (or downsize) me

How many rooms do you really need? The average American house size has grown to a whopping 2,479 square feet from 1,750 square feet in the late 1970s, according to figures from the National Association of Home Builders. A big house is more environmentally expensive than a smaller one, regardless of its green efficiencies. Big houses can mean more furniture, lighting and cleaning, and longer commutes to stores, work and school. Be realistic about what you need, go for good design and energy efficiency over maximum square footage, and you'll be creating a house that's proof against rising fuel and utility costs.

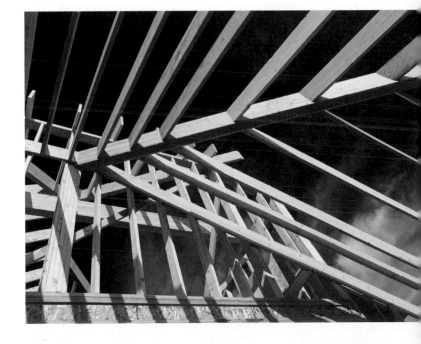

3

the lay of the land

Building is an opportunity to make sure your property has its own green footprint for future generations. Consider how your property relates to the natural topography of the site. Choosing and using a site resourcefully equals better energy efficiency. Understand the opportunities and limitations of your lot. Rectangular lots are generally more efficient in terms of land use, while sloping or steep blocks can require more excavation and fill and higher drainage costs.

Courtesy Environa Studio. Photo: Tim Wheeler Studios

4
face up to it

Courtesy Environa Studio. Photo: Tim Wheeler Studios

The orientation of your home has a huge impact on your utility costs. Take your lead from the environment and create a green floor plan. Because it receives the best of the day's sun, the south side is not only brighter but also warmer in winter, and perfect for living areas. The west side's afternoon sun makes it a better choice for bathrooms, garages and laundry rooms. The shady and cool north side is great for bedrooms in warmer climates, or for spare rooms. The east side, with its morning sun, is ideal for kitchens and bedrooms.

material world

5

Inverloch House in Victoria, by Solar Solutions Design.
Photo: Matthew Mallett

Prevent unnecessary landfill by recycling unwanted windows, bricks, timber, flooring and sheet metal in your renovation or build, or by selling them to second-hand building specialists. Eco-friendly building materials are growing in availability and affordability. Ask your builder and architect to avoid non-renewable resources in favor of green solutions, such as recycled timbers and steel reinforcements; engineered and plastic timber (great for decks); cement mixed with extenders; low- and zero-VOC paints, stains and finishes; and formaldehyde-free particleboard and plywood products.

insulate yourself

Metricon Home, courtesy of BlueScope Steel

Good insulation is fundamental, given that the average American home wastes up to 30 percent of energy used for heating and cooling as a result of cracks and gaps in the house's roof and walls. Insulation will help to regulate your home's internal temperature by reducing drafts and leakage of cool and warm air, keeping you warm in the winter and cool in the summer. Insulation also improves the effectiveness of your existing heating and cooling systems. Look for environmentally friendly and non-toxic insulation that will be good for the atmosphere in your home as well as the atmosphere of the planet.

6

7
green design

From igloos to adobes, humans have created climate-sensitive designs to perfectly complement their geographic location. In the early twentieth century, American architecture focused its efforts on adapting to regional materials and climates. But after World War II, popular design styles, such as modernism, became more generic and reliant on energy use to make homes comfortable, without regard for climate or location. When designing or renovating your home, engage professionals who share your green vision. Choose architects, builders, plumbers, engineers and contractors who think green too. The lowest quote is not always the best.

Courtesy Envirotecture, designers. Photo: Dick Clarke

8

passive attack

Banish the air conditioner. Fresh air and natural light are essential for your health, so invite them into your home, permanently, with passive design solutions that can reduce your reliance on electricity for heating and cooling. Create natural breezeways with the placement of doors and windows. Install skylights and enjoy free natural light for up to 14 hours a day. Venting skylights will improve airflow and air quality and prevent mold in bathrooms or laundry rooms.

9

window shopping

Because 25 to 35 percent of your home's energy for cooling or heating is lost through windows, improving their thermal performance will reduce energy costs and greenhouse gas emissions. Windows and exterior doors need to be energy efficient and smart. Consider the placement of windows to maximize natural light and create breezeways, and choose the right window for your environment. Double glazing is a good choice for freezing winters, since it entails less heat loss and less condensation. Louvered windows are great for catching the breeze on hot summer days.

Courtesy CplusC Design; Photo: Murray Fredericks

renovate or detonate

10

Courtesy CplusC Design. Photo: Adam Craven, Craven Images

Consider the benefits of renovating an old property against the drawbacks of building a new one. More than 70 percent of homes in America are over 20 years old—that's a lot of precious resources that can be renewed and updated. Older houses and apartments have a lot going for them: great location, more character, quality timber and established gardens. You can have the best of the old and the new by restoring original details, rethinking the floor plan for more modern living and installing energy- and water-efficient appliances.

cu green

Never discount the power of ideas when it comes to incorporating sustainable design into your home. The best results are often those that come from freethinking in the planning stages of your project's development. The 725-acre Palamanui sustainable community on Hawaii's Big Island is a prime example of how creativity can be shaped and channeled into green initiative.

In early 2008, Cornell University faculty members were approached by developers from Palamanui who were in search of fresh ideas from a group of unconstrained, energetic students. The result was CU Green: a coalition of 12 students and four permanent faculty members who

traveled to Hawaii on two occasions to brainstorm new methods for implementing green technologies in ways that hadn't been considered before.

"Generally I think it panned out pretty well," says Martha Bohm, a visiting lecturer for the Cornell Department of Architecture and a project adviser for CU Green. "The students were kind of the idea generators, and that's what the development was looking for—really creative thinking."

Some of the ideas generated included the use of a micro-grid power system to conserve energy, natural ventilation systems in place of air conditioning, and plug-in hybrid vehicles that could serve as a back-up source of energy for the community's electric grid. The students also favored the use of solar power controlled by Power Purchase Agreements, in which solar panels are installed on houses and maintained by the development which then sells the power back to the residents.

Martha says a sustainable community was a logical choice for Hawaii's Big Island, where residents pay huge costs for fuel and electricity is nearly three times as expensive as in California. Air conditioners are responsible for a large part of the electrical demand in Hawaii, and Martha says CU Green's greatest success came not from designing complicated cooling systems, but from making use of simple, low-tech alternatives instead.

Here are Martha's tips for limiting the need for air conditioning in your home.

1> Orientation is everything. Knowing how your home is positioned in relation to daily sunlight and wind flow will help you to use nature to your own benefit. Place windows strategically in order to create natural breezeways throughout your home, and consider how natural sunlight patterns throughout the day will affect the temperature in each room in your house.

2> Plant trees for shade. Thoughtfully planting trees—especially on the sunny south and west sides of the house—can reduce the need for artificial cooling significantly. If you live in a cooler climate, plant deciduous trees: they will drop their leaves in winter, so you're not receiving shade at a time when the sun's heat is beneficial.

3> Open up. Creating a home that is open to the outside will improve not only ventilation but also your quality of life. Making use of patios, courtyards and fountains can all help to create a more comfortable, livable habitat without having to rely on artificial cooling.

Photo: Helen Filatova/Shutterstock

green energy

11
warm up
with radiant heat

Nature uses it to warm us, and the Europeans have known about it for almost a century—radiant heating is one of the most energy-efficient and effective heating options available. By heating the floor, the ceiling or panels in the walls of your house, you can achieve a more natural and consistent warmth. Other advantages over traditional convection-based systems and reverse cycle air-conditioning are that you're warming your whole house (not just one or two rooms), dust circulation is reduced (great for allergy sufferers), and those fights over who's hogging the heater become a thing of the past.

you've got the power

W e are becoming increasingly aware of the running costs of everyday appliances like computers, televisions and refrigerators, but there are alternatives. Photovoltaic panels, which catch sunlight and transform it into energy, have been used on houses and commercial buildings for some time. Combine today's sophisticated paneling with the sun-drenched climates of many regions in the United States and it is possible to easily cover about 900 kilowatt hours (kWh) a year. While the average American family uses around 11,000 kWh annually, a household with energy-efficient appliances that makes smart decisions when using hot water, selecting lightbulbs and setting the thermostat could cut its energy use by up to 25 percent per year.

Courtesy Todae
(www.todae.com.au).
Photo: Lee Stone

13

leave the heat outside

Ever admired those quaint houses in Greece or Italy with their wooden shutters to keep out the sweltering midday sun? Well, there's a very good reason our Mediterranean friends have chosen this type of window cover. Outside awnings and blinds are much more efficient than their indoor equivalents, and can keep temperatures down by up to nine degrees. The iconic Southern porch, for example, is perfectly suited to the region's hot climate, shielding the exterior walls and rooms of the house from the harsh sun as well as allowing additional ventilation on hot summer evenings.

Courtesy Environa Studio.
Photo: Tim Wheeler Studios

14

clear the air

W e all see our homes as a refuge from the pollution and smog of the city, but the air within them can be more polluted than the air outdoors. Synthetic building materials, finishes and furnishings releasing outgas pollutants can harm your living areas. That sofa you sit on every night might be more detrimental to your health than running along your city's largest street in rush hour traffic. So do some homework on the materials you plan to use in your build or renovation and on those that are present in furnishings such as your carpets, couches, benches and cabinets.

Photo: Corbis Australia

15

nothing but hot water

Water heating is the third largest domestic energy expense on your utility bill. By replacing your old electric hot water system with more energy-efficient options, you can make your single biggest contribution to the environment. Think about switching to gas or solar systems for your hot water needs; and if you can't do that, install a ruthless shower-timer clock that cuts the hot water supply.

Courtesy Ian Kiernan, Westham Farmhouse

tanks for the free water

Walkerville House,
by Marc Dixon Architect.
Photo: Lucas Dawson

What could be more natural than harvesting rain that falls freely from the sky? Considering an average family uses 107,000 gallons of fresh water every year, half of which ends up down the toilet or in the garden, and you pay for it, installing a tank that can reduce this waste seems an obvious option. Tanks can be small or huge, above or below ground. They can even be flat so you can use them as a fence or under decks—the new generation of tanks is far from ugly. You can also do it yourself: set out buckets, bins and old toddler pools to collect small reservoirs (under the gutter is a great location) and use them on your plants when rain is scarce.

17

gray is the new blue

Don't just let all your waste water go down the drain. You can recycle it for non-potable (not for drinking) purposes with a readily available or more permanent gray water recycling system. Gray water—that is, all non-toilet household waste water from baths, showers, washing machines and sinks—can be redirected, filtered and treated, to varying degrees, for additional domestic use such as flushing the toilet and for outside use in gardens and landscaping. Consult your plumber and local board of health about the regulations and safety considerations of gray water use in your area.

recirculate hot water

18

It's frustrating, not to mention wasteful, waiting for hot water to flow through to the shower or faucet from the hot water tank. Waiting for just 30 seconds can result in almost 4 gallons of perfectly good water going down the drain. But a simple hot water recirculator can pump cold water silling in the pipes back to the tank until the hot water from the service is right at your faucet. A system can be connected to every outlet in the house with just one pump, which uses only a small amount of energy. Check with your local government to see whether you are eligible for a hot water recirculator rebate.

19

phase out inefficient light

Lighting represents a significant amount of household greenhouse gas emissions. By installing, where you can, energy-efficient alternatives such as compact fluorescent lamps (CFLs), you can significantly reduce the cost of your energy bills. CFLs use only 20 to 30 percent as much electricity as standard incandescent lightbulbs to produce the same amount of light, and they last up to 10 times longer. If every American replaced just one bulb in their home with a CFL, the savings to the environment and the economy would be enormous—over $600 million a year in energy costs and the elimination of greenhouse gases equivalent to the emissions of more than 800,000 cars.

Illustration: Marian Kyte

green
rewards

Photo: Marian Kyte

Sign up with your energy provider and play a real part in increasing the proportion of renewable energy (such as solar, wind, low-impact hydro and biomass) contributed to the national electricity grid. If you sign up with a government-accredited Green Power program, which will add a few extra dollars to your energy bill, your energy consumption can be converted into green credits so that you are literally investing in renewable electricity generation now and for future generations, replacing electricity generated from coal.

design solutions

Going green in your home is as much about the partnership with your design professional as it is about recycled materials and energy efficiency. Design Solutions in San Francisco, California, has been working with clients in residential and commercial interior design since 1986.

Although Design Solutions was founded as a traditional design and building firm, by 2005, the desire to take advantage of the burgeoning green movement in San Francisco and across the country convinced the company that green was the way to go—and their customers have responded. Today, Design Solutions offers clients a variety of sustainable alternatives, from recycled or sustainable solid surface materials and insulation to low-VOC cabinetry and LED (light-emitting diode) lighting.

Design Solution's Chris Connors says that although not all of their clients walk in with the intention of building or renovating sustainably, in many cases the company has standardized green products into basic design practices. "There are a lot of building materials now that are green that from a cost standpoint are no different," says Chris, who cites non-VOC paints and recycled denim insulation as two examples.

Chris says producing a successful sustainable result is often a matter of educating and negotiating with clients—finding out what they want and then sourcing comparable green alternatives. For example, a client who wants to use granite in countertops could substitute recycled glass or quartz.

Here are Chris's tips for incorporating sustainable interior design into home renovation or building.

1> Become an expert. The first step in choosing the green materials that work for you is to do your

Courtesy of Design Solutions

own research—and that starts with the Internet. By researching the green distributors and materials available in your area you can get an idea of the types of options that will be easily open to you. New products come out every day, and when it comes time to work with your design specialist, you can come to the table prepared.

2> Light up with LEDs. People sometimes forget the positive impact that LED lighting can have on their electricity bill—and the environment. New LED lights closely resemble incandescent bulbs, and new color spectrums make them an excellent alternative for every room in the house, including kitchens and bathrooms.

3> Don't get bamboozled. Though popular, not all bamboo is truly a sustainable material. In fact, many bamboo growers harvest their crop before it reaches maturity and suppliers may use urea-formaldehyde-based finishes and adhesives that are harmful to indoor air quality. Look for bamboo suppliers that can provide a "chain of custody" certificate to ensure their bamboo has been sustainably managed and harvested, and ensure that only water-based adhesives and finishes have been used.

21

room for room

Open-plan living solutions are popular with homeowners and builders alike, but think twice before you start to visualize how your family will coexist in a single, large space. The more you are able to separate individual zones in your home, especially transitional zones such as hallways, the easier it will be for you to keep these areas warm or cool when you have to rely on heaters or air conditioners.

Photo: Corbis Australia

22
watch your step

A British study found that about 40 percent of the environmental impact we create in building our homes comes from floor finishes. Although it has good insulating properties, carpeting can have a negative environmental impact and can exacerbate some health problems. Even pure wool carpet is often treated with chemicals to repel dirt and bugs. Try to source low-emission products, choose carpeting with low pile, and clean carpets regularly to discourage dust mites, using a vacuum with a high-efficiency particulate air (HEPA) system. If you opt for timber flooring, make sure you source it from sustainably managed forests or timber recycling specialists and choose low-toxic natural finishes, like tung oil or beeswax, over plastic floor finishes such as polyurethane.

light my fire

23

We love the romance of a wood-burning heater in winter, but inefficient wood stoves can send 15 to 30 grams of fine particle emissions into the air every hour. Check to make sure your wood stove complies with the EPA standard, that it is installed correctly, and that you are using it efficiently—closed, slow-combustion heaters are best. When building a fire in the fireplace, make sure that your wood is from a sustainable source. Wood smoke contains pollutants that can be harmful to your health, so use old and dry wood that burns cleanly and efficiently. Don't let your fire smolder overnight or use it to burn household trash, and have your chimney or flue professionally cleaned before each winter to maximize safety and efficiency.

Photo: Corbis Australia

beat the drafts

24

Your older home may have survived its share of renovations and design fads but the scars might still remain. When pulling up old carpet, beware of large gaps between the floor and the bottoms of doors that may have been ordered specifically or shortened to allow for carpeting. Revive the humble door snake and use rugs on gappy floorboards. Draft audit your home. Find green solutions to deal with other culprits such as gaps in walls and baseboards. Ensure that skylights, exhaust fans, downlights and chimneys are fitted professionally to minimize DIY-created drafts. Sealing your windows and doors properly can save you as much as 10 percent on your heating and cooling bills.

Courtesy Shiver Me Timbers

your biggest fan

25

C eiling fans are an attractive, quiet and inexpensive option for your cooling and heating needs. They work by circulating cool or warm air around the room and run at about 130 kilowatt hours (kWh) per year, compared with 500 kWh for evaporative cooling and up to 2,000 kWh a year for refrigerated air conditioning, which means you'll see a considerable difference in your energy bills, too. In the summer, place containers of ice under ceiling fans to create a cool breeze.

The Brisbane Sustainable House, courtesy of Environmental Protection Agency Queensland

26

don't stand by

Standby power can cost the average American household as much as $50 a year. When appliances are left on standby or "sleep" mode, they are not off. This wasted energy accounts for as much as 5 percent of all household electricity. The simplest way to save is to unplug them at the wall, or switch off the power strip. Just switching off electronics like your computer, television and CD player properly can decrease the more than 97 billion pounds of carbon dioxide produced by wasted standby energy every year. Ensure you have power strips with individual switches so you can isolate items you use less often, and use standby sparingly, to retain your settings on specific appliances.

27

let the music play

Your beloved stereo system does consume energy, but you can offset its footprint by choosing equipment from manufacturers that adhere to green standards, using recycled, non-hazardous construction materials and minimal packaging. You can also reduce waste by refusing to be lured into constant upgrades, and recycling whenever you can. It's been estimated that over five pounds of materials are used in making just one CD. Reduce the load by going digital and storing music on your computer or digital media player—and don't forget to turn off the standby mode.

Photo: Corbis Australia

star performers

W ith an average of 2.73 televisions per American household and the changes in tochnology over the last decade, your TV has become a major culprit in greenhouse gas emissions, behind refrigerators, dishwashers and water heating. When shopping for your next television, look for Energy Star and Energy Guide labels to help you make the greenest choice. Energy Star labels certify that a product has met the strict efficiency guidelines set by the U.S. Department of Energy and the U.S. Environmental Protection Agency.

Photo: Corbis Australia

29

window dressing

A dopt your grandmother's decorating tips and dress up your windows with curtains, blinds, shades, awnings and shutters to instantly improve the thermal properties of your house. These old-fashioned measures are not only back in style, but will minimize heat loss and drafts in winter and keep the hot sun out in summer—all further reducing your heating and cooling bills. Some things don't even cost you money: closing the curtains at sundown will stop heat from escaping through your windows, for instance, while keeping your house closed up on summer mornings will retain the cooler night air until a cooling breeze arrives in the evening.

Inverloch House in Victoria, by Solar Solutions Design.
Photo: Matthew Mallett

30

don't get sucked in

At least 80 percent of American households report having a vacuum cleaner, and the number of vacuums sold each year is increasing—which raises two questions: what are we doing with all those old cleaners, and how many vacuum bags go into landfills? Invest in a bagless cleaner to reduce waste, and choose a machine with a long warranty, so you're not adding it to landfill after just a few years. If it breaks, get it repaired. Clean filters and tubes regularly to maximize energy efficiency.

Photo: Corbis Australia

william mcdonough + partners

From the European headquarters of Nike in the Netherlands to the Adam Joseph Lewis Center for Environmental Studies at Oberlin College in Ohio to the sprawling Ford Truck Plant in Dearborn, Michigan, William McDonough + Partners is reshaping the way the world views green architecture and community design.

Founding partner William McDonough, a leader in the American sustainable design movement, heads a 50-person team of architects based in Charlottesville, Virginia, and San Francisco, California. The firm's philosophy centers on the idea of imitating the systems and principles of nature while considering the consequences of long-term design. In terms of their buildings, William McDonough + Partners works to eliminate waste, rely on renewable energy and honor diversity in their designs.

These key concepts are often referred to as being a part of William McDonough and Dr. Michael Braungart's "Cradle to Cradle" philosophy, which views human beings as part of a web of interconnected life, as opposed to the top rung of a ladder. In terms of building design, this means going further than the traditional standards of environmental efficiency, and making buildings more positive, as opposed to less wasteful or less polluting. William McDonough + Partners focuses on equipping

its projects with natural-day lighting and ventilation and providing access to views of the outside. More unusual features include the Ford Truck Plant's 10-acre vegetative roof and advanced rainwater-capture systems like that of the Bernheim Arboretum Visitor Center in Clermont, Kentucky, which uses decorative water features for natural beauty and, more practically, for on-site storm water management.

Here are some green building guidelines based on William McDonough + Partners' sustainable philosophies:

1> Get nurtured by nature. Make E. O. Wilson's theory of "biophilia," love of life or living systems, relevant in your own life. Embrace design that uses daylight to brighten rooms and allows inhabitants to use the sun's path to keep track of time during the day. And remember, most any view of the outdoors is a good one.

2> Have a plan (Part I). Every good design starts with sound building theory. For William McDonough + Partners, this means understanding the principles and goals of the people they are building for, and determining how those values will change over time. Before you go green, figure out what you want to achieve, and how that goal could evolve later on down the road.

3> Have a plan (Part II). Strategize. Understand the broad categories involved in any green architectural project. These include water, atmosphere, building site, materials and indoor air quality, and can be as specific as the food that will be eaten at a place or the method of mobility the people living there will utilize. Define your tactics for achieving results and develop a metric—that is, a system by which you will eventually measure your success or failure in each area.

31

green gourmet

Choosing new kitchen cabinets can present a real green challenge, but set yourself some basic rules—like sourcing materials locally, avoiding particleboard and fiberboard, and insisting on low-VOC cabinets. Countertop materials should be durable and water-resistant; stained concrete or indigenous stone are good options. Alternatively, source a recycled kitchen from a recycling center or online, and customize it to fit your needs. Mix and match with new products if you can't find exactly what you want second-hand.

32

fridge magnets

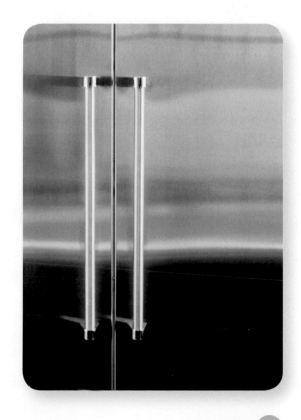

Refrigerators are a central part of our modern lives, but they are one of the biggest energy-consumers in the home. The average household refrigerator uses over 1,000 kilowatts (kWh) annually and a freezer uses around 800 kWh, generating over 2,000 pounds of CO_2 every year. Reduce consumption by improving your fridge's efficiency. Buy a machine with a good EnergyGuide rating, and don't buy one that's too big for your needs. Look for a refrigerator with automatic moisture control, as models that use an "anti-sweat" heater to reduce moisture require 5 to 10 percent more energy. Check the energy costs of your refrigerator and other appliances at the U.S. Department of Energy's Energy Savers site (www.energysavers.gov).

ye olde fridge

33

New Energy Star refrigerators use 40 percent less energy than conventional models made in 2001. If you're stuck with an older model for a little longer, however, congratulate yourself on not contributing to the local landfill and don't despair; there are a few simple things you can do to reduce your running costs and make your fridge more efficient. Keep the thermostat at an appropriate level so you're not wasting energy on freezing temperatures. If the fridge is near a window, draw the curtains to keep the sun off it. Clean the condenser coil (usually at the back of the machine) and make sure the seals on the doors are maintained. When you go on vacation, empty the refrigerator and turn it off, leaving the door slightly ajar.

Photo: Corbis Australia

the wash-up 34

Photo: Corbis Australia

An energy-efficient dishwasher typically uses one third less water than average models. Ensure your newer dishwasher isn't power hungry by checking its EnergyGuide label. If you're buying a new machine, look for an Energy Star model, which uses 40 percent less energy than the federal minimum standard for energy consumption. Since the majority of energy required to operate a dishwasher is devoted to heating water, choose a model that allows you to reduce the amount of hot water used. Opt out of the automatic air-drying mode and choose fast or economy cycles when you can. Only run the dishwasher when you have a full load.

35
in the oven

In America, neither gas nor electric ovens are labeled for consumers with an energy efficiency or EnergyGuide rating. Over 50 percent of the energy used by the oven is wasted—so consider alternatives like the microwave, an electric frying pan or a pressure cooker. Use your oven efficiently: ensure the seals are tight so that heat doesn't escape, and make sure the inside light works so you don't have to open the door too often to check on your roast or cake. Go green and avoid using toxic chemicals to clean your oven; alternatively, wipe it down with a soapy cloth when it's still warm, or fill a roasting tray with water and heat the oven on medium until the water is almost evaporated, then wipe it down and scrub stubborn spots with baking soda.

Photo: Corbis Australia

36
check what's cooking

As a rule of thumb, electric cooking uses at least twice as much energy as gas cooking. Electric induction cooktops, however, can be 25 percent more efficient than electric ranges, so consider this option when it's time to get a new one. Just putting lids on saucepans and simmering food gently rather than boiling vigorously can make a big impact. Remember that every 10 cups of water boiled off generates about half a pound of CO_2. When the weather permits it, make use of your grill for everyday meals, as cooking outside saves you from having to switch on lights and exhaust fans.

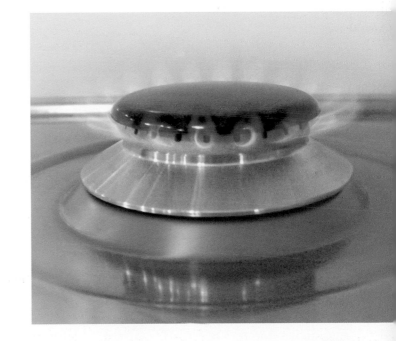

Photo: Marian Kyte

37
making waves

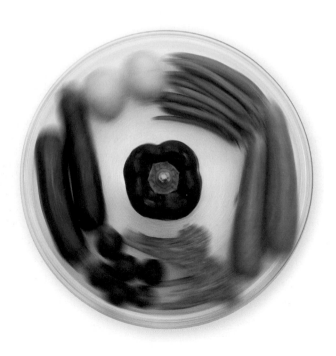

Microwave ovens are very efficient cooking appliances, largely because they use a relatively small amount of energy to achieve the required results. Choose an appliance with automatic sensor controls, a rotating turntable or stirrer fan and a computerized control panel to ensure precise cooking time and power. Keep in mind that regular maintenance on your microwave will help it last longer: a unit with a ruined waveguide cover (which may spark or burn) is expensive to fix because of the labor involved, and many machines are thrown out because of this. Ensuring this part is kept clean, according to the manufacturer's instructions, can extend the life of your appliance.

Photo: Marian Kyte

around the hood

38

W ithout an over-stove exhaust fan, it's likely that more than just a smattering of cooking fat will end up on your walls, ceiling and cupboards. Range hoods are great for ridding the air of grease, steam and odor—which can also reduce your repainting time—but they can be major energy eaters too. Choose a unit with fluorescent overhead lights, which last longer than traditional halogen ones, and ensure it's installed between two and two and a half feet above the stovetop for maximum efficiency. If you have a recirculating hood, as opposed to a ducted one that transfers air outside, make sure you clean the outside filter as well as the inside carbon filter regularly.

39

make a splash

Backsplashes are a popular choice to protect surfaces around sinks, stoves and countertops. But the traditional tiled backsplash—especially one with a tiny mosaic design—can be a nightmare to clean without resorting to chemical cleaning products. Explore other design options that will be easier to maintain—such as stainless steel, glass, laminate, larger tiles, marble or granite —and source second-hand or recycled materials when possible.

Courtesy Shiver Me Timbers

tools of the
trade

Ensure that everything in your kitchen can last a lifetime, avoiding unnecessary disposal in toxic landfills. Don't buy cheap tools that can bend, break or melt quickly. Invest in quality plates, serving dishes, baking equipment, pots and pans. Choose energy-efficient toasters, kettles and other appliances with long warranties. Wooden spoons can be rejuvenated and sanitized by a good boil in a pot of water. Invest in tight-grained wooden chopping boards from recycled or sustainable sources, and ensure their longevity by oiling them down with linseed or olive oil; instead of soaking them when knife cuts look deep, sand and oil them again.

tepeyac haven

No one should be denied the right to live in a non-toxic, sustainable structure. The new era of green building isn't just for the wealthy, and Catholic Charities Spokane's LEED gold-rated Tepeyac Haven housing project in Pasco, Washington, proves it. "LEED" refers to the standards set by the U.S. Green Building Council's Leadership in Energy and Environmental Design project.

The thinking behind Tepeyac Haven—a project of Zeck Butler Architects and the Beacon Development Group—was simple: build compact, energy-efficient structures whose sustainable attributes will pay for themselves over time. Savings over time was an especially important consideration for Catholic Charities Spokane, given that the 45 families residing in Tepeyac Haven were made up mostly of low-income farm workers from southeastern Washington's Tri-Cities area. The 44-unit project commenced in fall of 2007. "Within a month the thing was completely rented up and there were 120 people on the waiting list," says Dr. Robert McCann, Executive Director of Catholic Charities Spokane. "We instantly heard from the residents that this was the nicest place they'd ever lived in in their lives."

High density (15 units per acre) and efficiency were key concepts for Tepeyac Haven, both to reduce the development's environmental impact on the site itself, and to limit the cost of operations. Each unit came stocked with energy- and water-efficient appliances, and special attention was given to the development's landscaping, which, to conserve water, limited grass lawns and maximized the planting of drought-resistant shrubs. The site for the community was also important to increase livability; playgrounds for children were included in the development and Tepeyac Haven is located within walking distance of transit centers, stores, and schools. "We wanted to pick a location that made sense for the families," says Rob.

Courtesy of Zeck Butler Architects

Here are some of the things Rob learned from working on Tepeyac Haven.

1> Use the "justice lens." Choosing to make any green decision—whether in a community housing project or a bathroom renovation—is just in two ways. One, it is just for the environment, and two, it's just for the people who will be living in or utilizing the facility. Consider the justice lens when making green decisions in your own home.

2> Go for the gold. When you're talking green, initial costs may be high, but consider savings over the long term. Tepoyac Haven residents typically save $75-$100 per month on their utility bills, paying one third less than residents of similar, non-green housing communities. If you're going for LEED certification, consider that the higher the certification, the greater your savings could be.

3> Hang tough. The planning stage of green building can often be a daunting task. Discussing costs and green additions with your architect and contractor might be overwhelming, but do your research and stand your ground. Your investment will pay off in the end.

hot water systems

The production of hot water accounts for 13 percent of your electric bill, yet much of the energy is wasted by heat loss along pipes and tanks. You can minimize this loss by making sure your pipes are well insulated and your hot water unit is as close as possible to the place where you'll need it most. When choosing a new hot water system, get one that suits the needs of your family. The larger the system, the more heat it will lose and the less environmentally sound it will be. Avoid a system with a continuous pilot light, as powering a pilot light can release up to 450 pounds of CO_2 every six months. Try not to use hot water if you can use cold instead. Put a bucket in the shower to catch the cold water while you wait for the hot water to kick in, then use it on your potted plants or garden.

The Brisbane Sustainable House, courtesy of Environmental Protection Agency Queensland

waterworks

When choosing new fixtures and appliances for the bathroom, make sure you base your decision on the Environmental Protection Agency's WaterSense labeling system. A WaterSense label certifies that fixtures such as sink faucets, showerheads and toilets have been tested independently and certified to meet the EPA's criteria for performance and water-saving efficiency. Switching to a WaterSense-labeled bathroom sink faucet, for example, could save a household enough water to do 14 extra loads of laundry every year. For more information on the WaterSense program, go to www.epa.gov/watersense.

43

sink or swim

Don't let the faucet run, wasting our most precious resource, as part of your family's daily bathroom routine. Embrace the drain stop and use your sink as it was intended, by pouring in some water to wash your hands and face or to shave. Don't run the faucet when brushing your teeth—use a cup of water instead. Any of these simple ideas can save as much as one gallon of water a minute. Check that your drain stop is actually keeping water in the basin, as a leaky plug equals water wasted.

Courtesy CplusC Design. Photo: Murray Fredericks

44

recycled fittings

If you're updating your bathroom, give some thought to where new materials come from and the amount of energy used to produce them. There are lots of opportunities to source stylish, recycled and no-longer-wanted bathroom fittings rather than buying new, mass-produced, energy-intensive items. You can now get ceramic floor tiles made from recycled windshields, and retro basins and cast-iron bathtubs are readily available. Ensure that your cupboards and counters are derived from sustainable (rather than old-growth) materials. Check your telephone book or the Internet for recycling centers near you.

Photo: Corbis Australia

45

flushed away

Toilet flushing is responsible for about 30 percent of all water used by the average American household, which means that almost 44,000 gallons of quality water is flushed away per family every year. Upgrade to a high-efficiency or WaterSense-labeled toilet; without one, your single-flush unit can use 3.5 to 7 gallons in one flush, compared to a high-efficiency unit's 1.3 gallons or less. Check with your local government—you might even be able to switch to a waterless composting toilet. If you're stuck with your old model, place an inexpensive toilet dam (available from your hardware store) in the cistern to reduce the amount of water in each flush, or fill an empty soda bottle and place that in the cistern. Silent leaks can waste up to 200 gallons a day: drop a little food coloring into your tank and wait 15 minutes—if the color turns up in the toilet bowl, call your plumber.

reading matter

46

T hat two-ply designer toilet paper is indeed an expensive luxury. Amazingly, every ton of paper recycled saves 17 trees, 380 gallons of oil, 4,000 kilowatts of electricity, 3 cubic yards of landfill and 7,000 gallons of water. But recycled toilet paper makes up only a 2 percent share of the American market —much of the rest is made from plantation-grown or native forest trees. Make the switch to environmentally friendly toilet paper that is unbleached, chlorine-free and, if possible, 100 percent recycled.

light showers

47

Showers are the biggest resource-guzzlers in the home, but you can reduce this just by spending less time in the shower and using a timer. Replacing your old showerhead with a WaterSense-labeled or energy-efficient model can save you almost 3 gallons of water a minute. Take advantage of rebates and offers provided by your local government or water authority to make the switch. If one out of every 100 U.S. homes was retrofitted with water-efficient fixtures, we could avoid 80,000 tons of greenhouse gas emissions a year. Keep a bucket handy to collect gray water for use in the garden. Install a thermostat that keeps hot water at a usable temperature so no cold water is wasted.

Courtesy Interbath Australia

in the tub

M any Americans are simply getting used to showering instead of bathing, since a bath uses up to 70 gallons of water compared with a quick 10-gallon shower. But there are occasions when a bath is indispensable, such as for therapeutic reasons or when bathing a child. There are still a few things you can do to reduce water and energy consumption, though: bathe the kids together, insulate around the tub or install an in-line heater to keep the hot water hot so you won't have to constantly refill. After their bath, use the gray water for the garden, and make sure to get the kids involved.

Courtesy VELUX Australia

49

pick up the steam

It is essential to have an exhaust fan in your bathroom to reduce mold and other harmful contaminants, but that's no reason to forget the fan's negative environmental impact. Choose energy-saving exhaust fans that are thermostatically controlled and not connected to the light switch, so they turn on when the air temperature rises rather than when someone turns on the light. Clean the fan's filter once a month to ensure that it runs efficiently.

eco-accessorize

50

Make your washing area an eco-friendly haven with a few small changes to your bathroom basics. Toss out that PVC plastic shower curtain in favor of a natural alternative, such as 100 percent organic hemp or a heavy cotton one that doesn't cost the Earth (literally) in its production and is more naturally resistant to mold. Update your bathroom linen with organic cotton towels. Choose Earth-friendly soaps and cosmetic products. That plastic bathmat can go too, in favor of a washable cotton alternative. And don't forget to recycle bathroom containers for shampoo, conditioners, bubble bath, mouthwash and so on.

f.w. honerkamp co.

When it comes to choosing sustainable materials for your renovation or build, knowledge is power. This is especially important when choosing wood products for your home's cabinet and paneling work. The F. W. Honerkamp Company, located in the Bronx and Central Islip, NY, has been an established force in industrial woodworking since 1871.

Owner Jeff Honerkamp is a part of the fifth generation of Honerkamps, and has been interested in sustainability since his college years. It's a passion he later incorporated into his family's business; in the past five years, F. W. Honerkamp has become a leader in sustainability in industry, promoting green and renewable products and materials and educating the customers on the best ways to bring sustainable woodworking into their homes. In order to standardize the products they carry, Honerkamp looks for approval from third-party organizations such as the U.S. Green Building Council's Leader in Energy and Environmental Design (LEED) rating system and the Forest Stewardship Council's (FSC)

certification process. The company also has LEED-accredited professionals—including Jeff Honerkamp himself—to assist with customers' design concerns.

For Jeff Honerkamp, keeping his customers and other members of the woodworking community up to speed on sustainable wood products is a challenging process, though the information itself is relatively easy to explain.

"With wood products it's very simple," says Honerkamp. "It typically comes down to two different things: what adhesives or what things are in the wood, and where the wood comes from. That's it."

Here are Jeff's tips for bringing green woodworking into your home.

1> If you use composite wood products in your home, choose a third-party green-certified manufacturer that uses 100 % recycled wood or off-cast materials in its construction, and make sure that no urea-formaldehyde has been added during the manufacturing process. Look for water-based formaldehyde adhesives instead of urea-formaldehyde adhesives, which can cause harmful off-gassing.

2> Choose your finish wisely. Most custom cabinetwork requires sealants and finishes that can be potentially harmful to your indoor air quality and your family's health. Look for a low-VOC finish (typically water-based) as opposed to standard finishes that are higher in VOC's.

Photo: Marc Slingerland/Shutterstock

Courtesy of F.W. Honerkamp Co.

3> Know where your wood is coming from. Recognizing the difference between sustainable and non-sustainable wood is a lot like recognizing the difference between an organic and non-organic apple—you can't tell just by looking. Like an organic label, however, certification by the Forest Stewardship Council guarantees that your wood has been grown in a sustainable way and harvested from a sustainable location—as opposed to being clear cut from old-growth forests or harvested illegally.

sleeping green

51

step lightly

Since you spend more time in your bedroom than just about any other room in your house, don't ruin your pristine sleeping environment with volatile wall finishes or flooring. Opt for environmentally friendly paints. Choose solid wood, bamboo or recycled cork over vinyl flooring; natural-fiber rugs and mats that have not been treated with stain-resistant finishes are a good choice. Check the labeling on wood stains and polishes—anything containing wood-preservative chemicals in the active ingredients should be avoided.

a natural mattress

52

Synthetic materials used to make mattresses are not only energy-intensive to create, but they are not biodegradable and add to problem waste. They can also contain harmful glues, foams, pesticides and flame retardants. Natural fibers, however, have built-in protective properties, and readily absorb and release the moisture we emit while we sleep, so they're less likely to attract dust mites and bacteria. Look for products made from wool, organic cotton, latex and sustainably harvested wood. Air out natural-fiber mattresses in the sun—this will kill dust mites and other organisms.

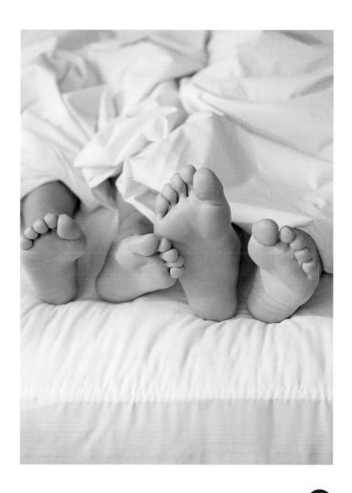

53
make your bed

Choose bed frames and bedroom furniture made from solid wood—either second-hand, reclaimed or from sustainable forests—or from sustainable materials such as bamboo. Plywood and particleboard consist of wood pieces held together with toxic glues; they can be less structurally sound than solid wood and may not last as long. Moreover, the source of plywood and particleboard is often untraceable, and the materials may come from old-growth forests—irreplaceable treasures that are being lost at a rate of 32 million acres a year. Support manufacturers who follow a green procurement policy or are certified by the Forest Stewardship Council (FSC).

between the sheets

54

Make the switch to organic sheet sets and sleep soundly. Conventional cotton is a water- and pesticide-intensive crop with significant environmental impacts throughout its production, manufacturing and processing. Support positive change in the industry by updating your linen cabinet with certified 100 percent cotton bedding that is unbleached, chemical-free and uses vegetable-based dyes. You can also get sheets made from hemp or bamboo, both of which require far less (if any) pesticides and herbicides to grow.

eco—air freshener

55

There is a simple way to dilute the pollutants that build up in your home—all you need to do is open the windows! The exchange rate—the rate at which the entire volume of indoor air is replaced by outdoor air—is .7 to 1 per hour for an average U.S. home. Keep your fresh air from windows really fresh with natural-fiber curtains and blinds. Organic cotton, wool, flax (linen) and hemp drapes are good insulators, naturally long-lasting and strong, and natural dyes are more fade-resistant than synthetic ones. Don't forget about the benefits of using houseplants throughout your home. From the bathroom to the living room, houseplants are nature's own air fresheners, removing toxins and unhealthy air pollutants and adding beauty to your living space.

Courtesy VELUX Australia

pillow talk

Your lungs and skin will breathe easier with natural-fiber (not synthetic) pillows, blankets and duvet covers. You can get pillows made from latex, organic cotton, wool and buckwheat; duvet covers made from wool as well as feather down; and beautiful cotton and wool blankets. Synthetic fibers are ineffective and encourage chemical production; they also trap air and water near the skin and make for uncomfortable sleeping—especially for small children. Don't forget to use healthy and sustainable duvet and pillow covers. You can wash pillows, bedspreads and blankets in your washing machine, adding a few drops of tea-tree oil to kill off dust mites; fluff and air them out in the sun to dry.

57

cotton on

We've all stayed in hotels where those polite signs in the bathroom suggest you hang up your towel so you can use it again in order to save precious water and energy. Many hotels claim around 70 percent of guests happily opt to do this, and thereby help to achieve at least a 5 percent reduction in utility costs. It's a good idea at home too. Fresh-air drying your towels each day will cut down on the number of times you need to wash them. You'll just need to train the teenagers in your household to pick up damp towels off the floor and hang them out to dry. Remember to buy only 100 percent organic cotton towels when you update your linen cabinet—they also make great gifts.

Courtesy Neco and EcoLinen

storage dreams

Good storage systems are the key to a nice, comfortable bedroom. Look for options made from recycled wood and steel—you need them to be strong so they'll last. You don't necessarily need custom-built storage: customize old furniture yourself by removing or adding shelves or drawers, and use your imagination when it comes to stackable crates and boxes that can be readjusted as your mood or circumstances change. Remember to store off-season clothes in breathable containers such as rattan trunks or muslin bags. Cedar chips with a couple of lavender-oil drops on them are a safer alternative to pesticides such as naphthalene moth-repellent products (mothballs).

59

changing rooms

Chances are, the home that suits you today won't suit you tomorrow. Although a majority of single-family houses in America have three bedrooms according to the U.S. Census Bureau, most houses are occupied by fewer than three people. A room that's a bedroom this year might not be one in the next. When you plan your home, make sure that a room that's a designated bedroom could also serve well as a study, library or guest room, or could even become part of another area of the house by demolishing a wall. Flexibility in your design can help reduce emissions and materials.

60

keeping your cool

The Brisbane Sustainable House, courtesy of Environmental Protection Agency Queensland

When building from scratch or renovating, remember that the bedroom is the room in the house where constant temperatures are most important. Passive design principles and a few smart decisions make this possible even in the hot summer months. Ensure that walls (especially those that face east and west) are insulated, while roof overhangs, awnings or vegetation can be used to shade the house. If it's more than one story, consider having your bedrooms on the ground floor. High ceilings allow hot air to rise above the living space and are a good place for ceiling fans, while vents over doorways and high-positioned windows enable hot air to escape.

if green furniture

The new generation of green-chic furniture has arrived, and there's no hemp in sight. Portland, Oregon's IF Green has been building sustainable furniture since 2005; producing one of a kind pieces that are all about crisp, cool lines, sustainable woods and the hardiness to stand the test of time.

For IF Green's master craftsman, Stephen Becker, furniture making is in the blood. Becker's own grandfather was a German furniture maker, and although Stephen's mid-century modern tables and chairs, benches, beds and cabinets are classics, they are far from traditional when it comes to sustainability. IF Green furniture uses only recycled, sustainable or Forest Stewardship Council–certified woods, natural fibers for upholstery, low-VOC adhesives and natural finishes, like water-based waxes and linseed oil. But don't expect IF Green to substitute practicality for green points. "We're trying to make our furniture something people shouldn't just buy because it's green," says IF Green co-owner Lisa Grove, who is also Stephen Becker's wife. "Which means working on design elements and, of course, making it as affordable as possible."

Lisa says that using products that are safe and non-toxic is important not just for the customers who purchase their furniture, but for the IF Green employees who work to create it. "The fact is, there are sustainable equivalents to every single aspect of the process," she says.

Lisa says safe, green furniture is sometimes the last thing that occurs to people—even those interested in incorporating sustainability into the home. Here are Lisa's tips for choosing green furniture for your home.

1> Be a "wood geek." There are plenty of simple but beautiful wood materials to choose from that can make a

Photo: Nicole Seawright

terrific addition to your home décor. Regional wood that was once used as floor or ceiling timbers or old school bleachers can be recycled in your home's furniture or woodwork (IF Green uses Portland's native Douglas fir wood). Palm wood also makes a good sustainable alternative, since many palm trees in the tropics are burned when they reach old age, resulting in both physical waste and air pollution.

2> Upholster wisely. Green furniture is more than just wood and finishes, it's fabric too. New green fabrics are gaining popularity and coming online at a rapid rate, resulting in plenty of choices for savvy consumers. Research sustainable fabric alternatives like "happy" or synthetic leather and organic cotton, or look for fabrics that have recycled fibers or fibers that can be composted when they are no longer useful.

3> Buy to last. Choose furniture that can last a lifetime, not just a design season. Select pieces for both strength and style. Look for durable wood materials and ask if your green furniture manufacturer strength-tests its pieces. When it comes to upholstery, pick fabrics that will be stylish even 10 years down the road.

61

breathe easy

In America, around 15 million tons of VOCs (volatile organic compounds) are released into the atmosphere annually, and paint contributes to this figure. Choose low- or zero-VOC paints, stains and primers rather than oil-based products. Manufacturers can provide material safety data sheets (MSDS) listing ingredients and their possible impact. If you're renovating a property built before 1970, test it for paint containing lead. Know the safety precautions you must take when working with lead-based paints by reading the free government pamphlet, "Lead Paint Safety: A Field Guide for Painting, Home Maintenance, and Renovation Work" (www.epa.gov). Research natural-fiber wallpapers, made from bamboo, sisal and straw.

Photo: Corbis Australia

62

roll on up

When repainting your home, get your painting estimates right to avoid having paint left over. This will not only cut down on the clutter in your garage or storage room, but it will also help the environment. Most hardware stores will be able to help you find out how much paint you need. Just make sure you measure all of the wall and ceiling spaces before you go to buy paint—and don't forget to subtract the window and door areas from the total figure.

63
classics

When it comes to buying furniture, one of the best things you can do for both your wallet and the environment is to buy pieces that will last forever. While it may seem a little steep to fork out all that money for your dream chair when your car payment is due, if you scrape and save for a little longer, you could afford timeless, quality pieces that won't end up in the landfill. Have your chairs and couches re-upholstered to update your interior and look for good second-hand options that can be restored to their former glory.

Photo: Corbis Australia

ethical style

More and more manufacturers of furnishings are making commitments to green procurement and manufacturing standards. When you are decorating your home, make it your business to know what you're buying and who you are buying from. Check the labels. Do your research and don't be afraid to ask too many questions to be sure your piece of furniture also has a green conscience. Only when you know where the materials came from, what they were treated with and who made them should you purchase the furniture. That way you can be sure you're not encouraging unethical or unsustainable production.

65
mood lighting

There's nothing more charming than dimming the lights to set the mood of a room, but for soft and non–fossil fuel lighting, try a few candles. Avoid spending money on over-packaged, artificially fragrant paraffin wax products that are derived from crude oil and emit carbon dioxide. Beautiful beeswax candles are much more environmentally friendly and are often available from a local supplier; plus you can add a personal touch with a few drops of essential oil—much more romantic. Pure soy wax candles are another healthy and natural alternative to paraffin wax. Soy wax is also easier to clean if spilled on fabric—with a little rubbing and some warm, soapy water, it's gone.

Courtesy Neco

refreshing air

66

Reduce the air pollution in your home by airing your house regularly. Choose environmentally friendly products to keep your home smelling fresh —why use artificial room deodorizers, especially those that require electricity, when you can open a window and use the sanitizing effects of sunshine and fresh air to reduce food smells and pet odor? You can also make your own chemical-free room deodorizers cheaply, by mixing together a teaspoon of baking soda and lemon juice, or by adding a few drops of your favorite essential oil to two cups of hot water in a reusable spray bottle.

Photo: Corbis Australia

67

safe home office

Americans send over 34 printer cartridges to the landfill every minute. That comes to 2,040 an hour and about 18 million per year. However, it's easy to reduce unnecessary waste by watching your printer use and taking the time to recycle your empty cartridges responsibly. Most companies will provide instructions right on the box on how to either send back or drop off a finished cartridge for recycling. In some cases, it is even possible to have ink and toner cartridges refilled. Reduce the risk of any laser printer emissions that may be harmful to your health by ensuring your home office is well ventilated to allow airborne particles to disperse.

reduce e-waste

More than 70 percent of discarded computers and monitors and more than 80 percent of TVs become e-waste and are dumped in landfills. Many of these items contain dangerous materials including mercury and cadmium, which can leach into landfill sites and eventually the water system. E-waste is growing at three times the normal waste rate—the fastest growing stream of waste in the western world. The response to this is: repair, reuse and recycle. Fix items that can be repaired, reuse working components and find specialist recyclers by reading the United States Environmental Protection Agency's page on eCycling (www.epa.gov/osw/conserve/materials/ecycling). Failing that, donate your old computer to charity or use a registered recycling facility (www.earth911.org).

69
the paperless home

E ver heard of the paperless office? You can make a difference in your home by applying green office principles to paper conservation. Scan and save documents electronically rather than in paper files. Opt to pay your bills and do your banking online. Research has shown that by switching to electronic bills, statements and payments, the average household can save 6.6 pounds of paper, avoid the use of 4.5 gallons of gasoline, and prevent the production of 171 pounds of greenhouse gas emissions every year. Use recycled paper in your home printer, always print on both sides, and use old paper for scrap and note taking.

Photo: Corbis Australia

non-stationary stationery

Everyone uses pens, pencils, pins and packaging at some time, and you can make the most out of these products by thinking about them as permanent, not disposable, items. "Disposable" pens are not disposable—more than 10 billion plastic pens end up in landfills around the world every year. Bring some style back into your note writing by investing in a good-quality refillable pen (where you only need to buy ink). Reuse envelopes and packaging and choose refillable items whenever you can, including tape dispensers and pencils. When buying stationery, think recycled: you can purchase pencil cases made from car tires; rulers and personal planners made from old juice cartons; pencils made from old denim and post-consumer paper money; and mouse pads made from recycled circuit boards.

Photo: Marian Kyte

whole foods market

Few things have as much direct impact on our well-being as the food we put in our bodies. Whole Foods Market believes that we have a right to expect more from our food, and to that end, they provide organic foods that are sustainably and responsibly produced, and that nourish both their customers and the environment in the process.

Whole Foods Market was first established by the merging of two natural-food stores in Austin, Texas, in 1980. Today, Whole Foods has more than 270 stores worldwide, and while it may seem as if the company has come a long way from its humble roots, that success has only helped it share its philosophy of healthy, sustainable living with a broader audience. "We've been there from the forefront, really trying to work and nurture the organic industry and the environment at the same time," says Margaret

Wittenberg, Vice President of Global Quality Standards and Public Affairs for Whole Foods.

"People had seen how after [WWII] the increase of pesticide and herbicide use in agriculture… [brought] the promise of cheaper food, and more crops, and all that came with a price," says Margaret. And while the company believes strongly that organic produce is one method of protecting and encouraging healthy living for their customers,

Courtesy of Whole Foods Market

Margaret says it's just the tip of the iceberg when it comes to the company's sustainable initiatives. Because it began as a community store, Whole Foods believes in limiting its carbon footprint and in nurturing local growers by supporting their agriculture and offering $10 million in low-interest loans every year to help them get established. The store also sets a strong environmental example by eliminating plastic grocery bags and attempting to recycle and compost its own materials and reduce waste wherever it can. "It's kind of what we live and breathe," says Margaret. "It's what we do."

Here are some savvy shopping tips based on Whole Foods' sustainable food initiatives.

1> Fish around. Our global fish stocks are diminishing at a rapid rate. Be a responsible seafood shopper and do your research. Know where your local market gets its seafood and how its fish are fed and harvested. Aquaculture fish farming, for instance, is the more ecologically sound method of managing fish populations.

2> Buy local. Locally grown food is beneficial for several reasons. It supports local farmers and local economies, helps to preserve regional character, and helps to limit the carbon emissions and preservatives associated with shipping foods. Plus, the freshness of locally grown produce can't be beat!

3> Meat the challenge. As meat-eaters, we have a responsibility to ensure that the animals we eat have been raised and treated humanely. Only buy from meat producers who can guarantee that their animals were not fed animal by-products, hormones or antibiotics. Soon, humane meat products will be easily detectable when they are certified with an "Animal Compassionate" label.

care for your clothes

A staggering 3.5 billion wire clothes hangers end up in U.S. landfills every year. Don't just throw your hangers out when you bring home the dry cleaning. Why not take them back to your dry cleaner for reuse, give them to your local thrift shop or even save them for craft projects? If you're in the market for new hangers, seek out eco-friendly, recyclable alternatives made from recycled paper or cardboard. Buy long-lasting wooden hangers from recycled or sustainable sources, or old ones with a bit of character from second-hand stores.

Photo: Corbis Australia

pressing issues

If you feel that ironing is a waste of time and energy, try avoiding it altogether with some simple tricks. Use the cold water cycle on your washing machine (it's heat that causes the wrinkles in the first place); wash light clothing in one cycle, heavy in another (so the weight doesn't squish your delicates); bring laundry in from the line while it's still damp, so the sun doesn't dry wrinkles into the clothes; and hang clothes on hangers, so they retain their shape while drying. When you do iron, it makes better energy sense to iron in loads rather than piece by piece, and avoid wasted energy by not overheating the iron.

Photo: Corbis Australia

73

the natural finish

Contrary to all those advertisements, you don't really need a different cleaning spray for every piece of furniture. Not only does producing and packaging these cleansers place extra strain on the environment, but also the mineral oils and chemicals they contain are a major contributor to indoor air pollution. Instead, clean furniture with a damp cloth. If your furniture is made from untreated timber, you can use maintenance products made from natural oils and beeswax, which also have a beautiful smell.

Photo: Corbis Australia

carpet cleaners

It's important for your health to keep your carpets clean, as they can trap dirt and mold and harbor dust mites, but make that cleaning a more environmentally friendly process. Commercial carpet cleaning generates wastewater that includes chemicals and dirt from the cleaning process. Clean your carpets with environmentally safe chemicals and natural alternatives (like salt for red wine stains) and employ green-aware carpet-cleaning professionals. Place small, washable rugs made of untreated natural fibers over your carpets, or check out other flooring materials such as bamboo squares.

75

wipe out chemicals

More than 300 man-made chemicals can now be found in our bodies that people weren't exposed to just three generations ago, but we still aren't sure what effect these chemicals are having. Reduce possible exposure to any harmful chemicals in your everyday cleaning by using eco-friendly ingredients and methods. Your whole bathroom can be cleaned with a bit of baking soda, some vinegar and that old staple—elbow grease. Substitute natural germ fighters such as tea-tree or eucalyptus oil for bleach and chlorine, and reduce the amount of cleaning needed by regularly wiping down wet surfaces so that grime doesn't build up. Always choose detergent that is biodegradable with a low-phosphorus content.

The Brisbane Sustainable House, courtesy of Environmental Protection Agency Queensland

shine your kitchen

76

You simply don't need to buy a lot of harmful chemical products in wasteful packaging to clean your kitchen and utensils. There are books on natural cleaning—browse your local bookstore to find solutions such as the following. When your silver gets tarnished, soak it in a solution of hot water, baking soda and a little dish soap. For copper, try polishing with equal parts of ketchup and Worcestershire sauce. Brass comes clean with equal parts of vinegar and salt. Rub baking soda into tea and coffee stains, and clean countertops and sinks with a microfiber scrub and a bit of baking soda.

Photo: Corbis Australia

scrub up

77

Many soap and body-cleansing products contain large amounts of synthetic materials, which can be harmful to the environment and your skin and contain possible pollutants, allergens and irritants. There are now plenty of products that use organic or all-natural ingredients, or you can make your own. The Internet is a great source of homemade recipes using readily available ingredients. You could even collect soap bar remnants in a container and add water to make liquid soap.

Photo: Corbis Australia

78

ease the load

Front-loading washing machines use less water than top loaders, which can mean 40 percent less water, a savings of 7,000 gallons a year for a typical household. Front-loaders also need less detergent and consume less electricity. Look for Energy Star–labeled models and always wash with a full load: choose a detergent that is biodegradable with low phosphorus content. Also, heating the water for a hot wash generates almost 9 pounds of greenhouse gas emissions, but you'll generate less than 11 ounces of greenhouse gas per wash using cold water—so switch to cold washing and reduce water heating and greenhouse gases by more than 90 percent.

79

it's a wrap

In this day and age, plastic containers are available in so many shapes and sizes there should be little need to add waste to the landfill by using plastic wrap or paper or plastic bags. Get sandwich-shaped containers for lunches and take reusable containers to takeout restaurants and delis. When refrigerating leftovers, transfer them to containers with lids, or cover bowls with a plate rather than using foil or plastic wrap.

Photo: Corbis Australia

that sinking feeling

80

Check all the faucets around your house. New WaterSense faucets mean less water waste—and you can get flow restrictors or aerators fitted to existing taps (aerators lower the flow by adding air to the stream). Check for leaks and replace washers regularly—a tap that leaks at a rate of one drop per second can waste more than 3,000 gallons of water each year! Keep drain plugs in the sink when washing fruit and veggies, and use the water later for rinsing dishes or watering your indoor plants.

pleasant hill

Though they had always considered themselves to be environmentally conscious, building the first home rated "silver" under the U.S. Green Building Council's Leader in Energy and Environmental Design system was not Mort and Evelyn Panish's original intent. In fact, it was the Panishes' two sons who encouraged them to go green with their Freeport, Maine, home, aptly named Pleasant Hill. "You might say we learned from our children," says Mort.

Fortunately for Mort and Evelyn, Freeport, Maine, is also the home of Taggart Construction, an architectural firm that has been specializing in sustainable residential design since 1994. "Really from the start, our mission was to design energy-efficient, environmentally friendly, occupant-healthy buildings," says Peter Taggart, owner and president of Taggart Construction. Although Peter and the Panishes hadn't originally set out to achieve LEED certification, Peter decided to investigate LEED midway through the process. "It was an opportunity for us to get some verification of the work that we'd been promoting ourselves as for years," he says. In the end, only the addition of a garage exhaust fan and carbon monoxide detectors in the home were needed to achieve LEED silver based on Taggart Construction's original design.

Located on a hill, the Panish home makes use of its prominent southern slope for solar gain, along with a three-kilowatt photovoltaic array on the roof, high-efficiency appliances and a tiled-floor radiant heating system. But for the Panishes, what makes the biggest difference is their home freedom from drafts and its consistent climate control.

"It's extremely comfortable," says Mort. "It doesn't matter how cold it is outside, the temperature in the house remains rock-steady."

Here are some green home building tips learned by Mort and Peter during Pleasant Hill's construction process.

Courtesy of Taggart Construction

1> Don't get washed up. "Green washing" refers to the tendency of companies in the building or construction industry to falsely market themselves as green or exaggerate their green attributes. Do your research and ask your design professional before committing to a "green" company that may not be the real deal.

2> Meet often. When working with a green construction and design team, set aside a weekly time for a face-to-face progress report. This will not only help to have your questions and concerns addressed, but also keep you apprised of what materials are going into your home and how your budget is being spent.

3> Be a material girl (or boy). Materials that go into your home's envelope will make a daily impact on your family's health and well-being. Even natural materials like pine wood can have harmful effects if a member of the family has pine-tree allergies. Alert your builder to any chemical sensitivities or allergies in your family, and make sure he or she keeps you apprised of all materials that are being used.

81
going native

Many American gardens feature introduced plants that aren't suited to the climate. Some may require a lot of water, soil additives and pesticides to really thrive; worse, others may escape from your garden and become invasive pests. Growing your region's native plants is a great way to ensure minimum water usage and maximum gardening success because these plants are meant to be there. Ask your local nature center for a list of native plants and restore the natural habitat of your suburb, starting with your own backyard. Or ask your plant nursery for advice on plant types from similar climates that will work well.

compost

Americans toss out more than 43,000 tons of food a day—that's 15 million tons of food every year. But recycling your organic and green waste literally gives you something for nothing: 100 percent organic fertilizer for your garden that will improve your soil's water retention and the vitality of your plants by providing much-needed nutrients like phosphorus and nitrogen. Large gardens can keep great compost piles going; if you have a courtyard garden or are in an apartment, source a system suitable for indoors.

Photo: Corbis Australia

worm your way in

83

While your average garden worm won't quite do the job, you can establish a thriving worm farm in an unused corner with a few stackable, modified crates, some compost and newspaper and some brown-nose worms or red worms from your local nursery. If well maintained, your farm's population will double every two or three months. The basic idea is that worms dig upward to find fresh food scraps, leaving behind debris that makes incomparable fertilizer for the vegetable garden. You can inquire at your local garden shop for information or search the Internet for starter kits.

Photo: Corbis Australia

chicken little
84

If you have the space, keeping a couple of chickens in your backyard provides a great source of fresh, free-range eggs—letting the chickens out to roam and forage for a few hours a day, you will help keep the pests in your garden at bay. Each chicken also generates up to 45 pounds of superb organic fertilizer every year. Grab some two-by-fours and a roll of—you guessed it—chicken wire, and spend a weekend with the kids building the family a little chicken run. There are plenty of plans on the Web, but you should check with your local government for regulations and advice about keeping chickens.

85

mulch it up

Reduce erosion and the amount of water your garden needs by mulching regularly. Adding a layer of organic material to your garden beds reduces evaporation by up to 70 percent, reduces weeds and improves the quality of your soil, making your garden even healthier. Choose 100 percent certified organic mulch over other options, or save some money by using your own grass clippings, fallen leaves or shredded newspaper. You'll need enough to cover the soil in a layer two to three inches thick. Ask your local government or community waste disposal facility about mulch deliveries.

Courtesy Sustainable Gardening Australia

Photo: Corbis Australia

edible garden

Agriculture is a big consumer of fossil fuels. The fruit and vegetables in the supermarket may have traveled thousands of miles to make it to your table, not to mention the associated costs of packaging and storage. Support local market gardeners and suppliers or revive your own vegetable garden. If your climate permits, plant some fruit trees to source your own truly organic produce. From a large-scale vegetable garden to herbs grown in pots on a balcony, you can grow small amounts of produce to supplement your needs. There's plenty of organic gardening advice on the Web, and check your town for local gardening groups you can join.

87

friendly plants

In the wild, many varieties of plants grow in the same location, sharing their resources to help each other thrive. Use the principles of companion planting and let nature do some of the hard work, helping your garden thrive and stay pest-free. Using the right combinations of plants, you can naturally control pests by attracting beneficial insects and discouraging harmful ones. Insects, after all, are natural recyclers, pruners and composting machines.

Photo: Corbis Australia

bird life

88

Birds are great caterpillar and bug eaters, and making your garden a safe and pleasant habitat for them is easy. Provide fresh water by installing a birdbath or pond, and plant indigenous trees and shrubs to give birds natural food sources, such as nectar. Build appropriate nesting boxes out of recycled materials or find them at your local garden or hardware store and you'll be doing your bit for biodiversity. Use some old netting or chicken wire to keep birds away from seedlings and fruit. Avoid the use of garden sprays and poisons such as snail bait, and don't let your cat (or your neighbor's) roam around the garden.

89

be water-wise

The days of hosing your garden for hours on end or soaking the lawn with the sprinkler are truly over, and with good reason. Save water and reduce watering bills by creating and maintaining a water-wise garden. One of the best investments you can make is a drip irrigation system that delivers water directly to the roots of the plants. Design your water-wise garden with plants suited to your soil and weather conditions. Regularly check your outdoor spigots, pipes and plumbing fixtures for leaks. One dripping faucet can waste 250 gallons a month.

Photo Corbis Australia

pest control

A void the overuse of synthetic chemical fertilizers and pesticides in the care of your garden. They don't discriminate between beneficial insects and pests; moreover, they can be harmful to you and your garden's wildlife, and ultimately can contaminate our waterways. By using some commercially available pesticides you may also contribute to any adverse environmental impacts of the manufacturing process. Try organic pest remedies, instead. For example, mix one tablespoon of vegetable oil with three cloves of crushed garlic and leave it to soak overnight; the next morning, strain the mixture and add it to four cups of water and a teaspoon of liquid soap, and pour the mixture into a spray bottle. Rinse all produce, homegrown or store-bought, before use.

Photo: Marian Kyte

usgbc

How we build the structures we live and work in affects more than just the environment—green buildings can have a positive, long-range impact on human health and productivity. Since 1993, the U.S. Green Building Council (USGBC) has been setting the standard for sustainable building and design in the United States.

Today, the USGBC has more than 17,000 organizational members and 78 chapters and affiliates across the United States, and its LEED (Leader in Energy and Environmental Design) rating system is regarded by most as the foremost standard for building green in America. Professionals from every facet of the building industry—from architects and engineers to landscape architects, interior designers and public officials—can obtain LEED accreditation in their field, and almost

25,000 LEED-registered and -certified projects have been or are being built around the world.

"I think what we're most proud of is the fact that the U.S. Green Building Council has been able to bridge the gap between environmentalism and business," says Rick Fedrizzi, president, CEO and founding chairman of the USGBC. "We want to help businesses and corporations understand that they can have solid bottom-line performance and innovative business strategies while at the same time protecting the environment."

The payback from these buildings is twofold, since studies suggest that the lack of toxic materials, improved indoor air quality and natural lighting found in LEED-rated buildings can have far-reaching benefits for the people who occupy these structures—from students who perform better on tests to hospital patients who recover more quickly from their ailments.

Here are Rick's suggestions for using LEED principles to improve the efficiency of your home and the quality of your daily life.

1> Surround yourself in green. Embracing biophilia, or the innate need for humans to feel connected to nature and to their surroundings, can be beneficial to your health. Place potted plants around your home and patio, and if you have the space, plant and maintain a garden outside. Use small indoor and outdoor fountains to provide the aesthetically pleasing sounds of running water, and use wood instead of synthetic materials where you can in your home décor.

2> Rewind the clock. Sometimes futuristic thinking is the antithesis of green. Consider tailoring the operation of your home to the standards of your grandparents, rather than the Jetson family. Forgo the air conditioner when an open window will suffice, install a compost pile to turn food waste into fertilizer for next season's garden and make unattended running water a thing of the past.

3> Embrace what nature gives for free. Heat, daylight and water can all be put to good use in your home. Collect rainwater in barrels for use on your gardens and use thermal mass (such as stone or tiled flooring) to absorb the sun's rays and radiate back free, renewable heat for hours.

91

a cool house

One of the most efficient ways to keep your house cool inside isn't even in the house! Trees and plants that provide shade can significantly reduce the heat that penetrates your home, with direct sunlight generating as much heat as a single-bar radiator over each square yard of surface. Thoughtfully planting some trees and shrubs around your home is a long-term way of reducing the intensity of the sun on the eastern and western sides of your property. The effective use of deciduous trees can enable you to enjoy the warmth of the sun in winter and protective shade in summer.

Courtesy Environa Studio. Photo: Tim Wheeler Studios

outdoor lighting

92

We love our backyards, and outdoor lighting is great for extending the enjoyment of your garden into the evening, but it can also be a huge energy eater. Check out solar-powered lighting, which will charge during the day and emit a lovely glow at night. If you're after a more powerful lighting source, install timers or sensors so that lights aren't left on overnight. Don't over-light areas; instead, install switches to control individual lights.

The Brisbane Sustainable House, courtesy of Environmental Protection Agency Queensland

93

pave the way

Why not turn that high-maintenance lawn into a low-maintenance paved area? Did you know that using a powerful mower to cut the grass for an hour causes the same amount of pollution as driving your car for over 100 miles? Rethink the expanse of your lawn. There's plenty of paving options available, including recycled products, and paving means less watering, less mowing and less general maintenance. Scatter a few potted plants around your paved area to keep the garden spirit alive.

Photo: Corbis Australia

94

over the fence

When thinking about a new fence, go for long-lasting, recycled and environmentally friendly treatments. Reusable fence posts and rails can be purchased from recycling facilities and second-hand building suppliers. Buy new materials that are sustainable, such as bamboo and plantation timber. Wire fences are also available second-hand or constructed from recycled metals. If you're just after a simple garden or pet fence, recycled plastic fencing is available from hardware stores. Prolong the life of your fence by protecting it with non-toxic pest and rust deterrents.

Photo: Corbis Australia

95

cover the pool

Water restrictions are becoming a common occurrence in several regions in the United States, often affecting how and when you can fill or top up pools and spas. If you own a pool, ensure that your pool filter is the right size and is working correctly, so it's not consuming energy unnecessarily. Maintain the surfaces and check for leaks, which can waste up to 100,000 gallons of water a year. Invest in a pool cover or solar pool blanket to save on heating and cooling costs and reduce water evaporation.

playtime

You don't need to entertain the kids with energy-intensive, mass-produced play equipment. Use old tires (which are an environmental nightmare to recycle) for swings; old, untreated timber for sandboxes; and recycled wood and furniture for tree houses. You can also buy second-hand playground equipment and gym mats made from recycled materials.

Photo: Marian Kyte

out of sight
out of mind

97

There is considerable debate about the environmental impact of PVC pipes commonly used in domestic plumbing, particularly their end-of-life disposal and the associated issues of landfill, incineration and recycling. Weigh the benefits of polyvinyl chloride versus PVC-free alternatives like HDPE (high-density polyethylene) for your plumbing needs. HDPE pipes are readily available and are easy to use for the environmentally aware builder, and HDPE has a higher recycling rate to recommend it.

The Brisbane Sustainable House, courtesy of Environmental Protection Agency Queensland

recycling rules

98

It has never been easier to recycle, and about 31 percent of Americans regularly participate in recycling. Support your community's initiatives to ensure we have a trash-free environment. Your local or county government can provide information on recycling programs in your community, including procedures for the safe disposal of household chemicals and larger household trash items. Create a recycling center in your own home with a few well-placed containers: one for compost, one for recyclables, one for materials for the kids' craft projects and one for trash. Dedicate an area in the kitchen for the trash bins; it doesn't have to be huge, just big enough to suit your needs.

Photo: Corbis Australia

99

green clean your car

Cleaning the car is a weekend ritual that shouldn't be forgotten! Avoid chemical-based car polishes and detergents. There are new, waterless cleaning products on the market, or you can adapt your household green cleaning kit to your car (using baking soda to deodorize your car's carpets, for example). Don't use the hose—using a hose in hand for an hour can waste around 375 gallons of precious water. Wash your car with a bucket of water on the grass, so that your lawn can benefit from the extra water. And there's a good argument for letting the pros do it for you, if you choose a carwash that uses biodegradable, non-toxic detergents and reuses wash water several times before disposing of it.

Photo: Marian Kyte

100
shine on

Make room for a clothesline around your house, preferably in a sunny and windy spot. Using a machine to dry your clothes four times a week is expensive and produces more than 400 pounds of CO_2. Yet most people have access to a free dryer in their own backyard: the sun. Clotheslines don't have to be ugly and obvious—there are many options that fold down and away. Unlike many other appliances, clothes dryers are not rated by the Energy Star labeling system, since most current dryer models use the same amount of energy. If you do need a dryer, make sure it runs properly by cleaning the lint trap regularly and checking that the machine has adequate airflow. Set up a clothesline under a covered area to dry your laundry on rainy days.

jordan meadows

For most of us, achieving environmental excellence in the design and operation of our homes seems like a daunting, if not impossible task. Not so for Tom Snyder and his wife Lee, of Gorham, Maine, who decided to downsize their home after their children were grown to make it as environmentally efficient as their budget would allow. The result: a "gold" rating from the U.S. Green Building Council's Leader in Energy and Environmental Design system.

To build their 1,700-square-foot Jordan Meadows dream house, Tom and Lee turned to Freeport, Maine's Taggart Construction team, who were, at that time, working on the Pleasant Hill LEED-silver home in Freeport. "In order for a house to be sustainable, it's not just about one thing," says Peter Taggart, owner and president of Taggart Construction, "It's really about the whole house as a system and everything working together...." Taggart Construction worked closely with Tom and Lee to develop a whole-house plan, including everything from the materials in the home's envelope to light fixtures and faucets.

"Each time there was a decision to be made, we considered the cost/benefit of using the LEED guidelines and we chose to implement the highest level of performance that we could afford," says Tom. To that end, Taggart Construction used locally harvested and milled siding materials; recycled, Dense-Pac cellulose insulation in the walls; and high-efficiency plumbing fixtures. Jordan Meadows also uses a radiant heating system in its floors, which are made of four-foot squares of black concrete that look like slate and contain radiant tubing. Peter says the floor acts like a battery, absorbing the heat from the sun through windows, heating the floor and then radiating the heat out into the rest of the home.

Here are some green home building tips learned by Peter and Tom during Jordan Meadows' construction.

Photos: Randolph Ashey

CASE STUDY

1> Budget your budget. Once you've established the cost of the entire project, keep in mind that there will be decisions to make as the construction process goes on, mostly in the case of building materials. If your budget is pretty tight, remember that your primary obligation is to the home's envelope; never skimp on quality materials that will keep your home warm, dry and maintenance-free in the future.

2> Take to the trees. When building a green home, site preservation is especially important. Building around established trees instead of cutting is an option that will offer your home privacy, shade your home from the summer sun and help keep it warm in the fall and winter.

3> Out of sight, in your mind. Green isn't just about the solar panels that you see every day. Keep in mind that some of the greenest and most important materials in a house—such as recycled, dense-packed cellulose insulation, Forest Stewardship Council-certified woods and concrete radiant heating systems—may not be as readily apparent.

resources

websites

Green building resources & industry organizations	Efficient Windows Collaborative	www.efficientwindows.org
	National Association of Home Buildersw	www.nahb.org
	Build it Green	www.builditgreen.org
	Global Green USA	www.globalgreen.org
	World Green Building Council	www.worldgbc.org
	Green Roundtable	www.greenroundtable.org
	Architects/Designers/Planner for Social Responsibility	www.adpsr.org
	USDA, Home Conservation Advice	www.nrcs.gov/feature/backyard
	No. American Insulation Manuf.'s Assoc.	www.naima.org
	U.S. Green Building Council	www.usgbc.org
	Sustainable Buildings Industry Council	www.sbicouncil.org
	The Green Building Initiative	www.thegbi.org/home.asp
	The Energy and Environmental Building Assoc.	www.eeba.org
	DesignGuide.com	www.designguide.com
	Building Green.com	www.buildinggreen.com
	ClimateBiz	www.climatebiz.com
Green homes	Build it Solar, Renewable Site for Do It Yourselfers	www.builditsolar.com
	WormMainea	www.wormmainea.com
	Healthy House Institute	www.healthyhouseinstitute.com
	Green Home Guide	www.greenhomeguide.org
	Green Building Pages	www.greenbuildingpages.com
	Regreen	www.regreenprogram.org
	Good to be Green	www.goodtobegreen.com
	Build-e, Eco-Friendly Houses	www.build-e.com
	Certified Forests Products Council	www.certifiedwood.org
	Environmentally Construction Outfitters	www.environproducts.com
	Environmental Home Center	www.environmentalhomecenter.com
	Sustainable Home	www.sustainablehomemag.com/
Green gardens	Coalition Against the Misuse of Pesticides	www.beyondpesticides.org
	Eco-Friendly Plants	www.eartheasy.com
	EPA, Indigenous Plants Landscaping	www.epa.gov/greenacres
	Gardens Alive!	www.gardensalive.com
	Gardener's Supply	www.gardeners.com
Tools	Carbon Independent	www.carbonindependent.org
	Carbon Footprint	www.carbonfund.org
	Carbon Calculator	www.americanforests.org
	Ecological Footprint	www.myfootprint.org
	Energy Calculators	www.eere.energy.gov/consumer/ calculators

Green labeling	Electronic Product Environmental Assessment Tool	www.epeat.net
	Green Seal Cert. Cleaning and Paper Products	www.greenseal.org
	Ecolabels	www.eco-labels.org
	Global Ecolabeling Network	www.gen.gr.jp
	Transfair USA	www.transfairusa.org
Water saving	Greywater	www.greywater.com
	Soil Water Conservation Society	www.swcs.org
	Water-Saving Tips	www.h2ouse.net
	Water Use It Wisely	www.wateruseitwisely.com
Green power	Get Energy Active	www.getenergyactive.org
	Sustainable Energy Coalition	www.sustainableenergycoalition.org/
	Alliance to Save Energy	www.goodtobegreen.com/
	Energy Information Administration	tonto.eia.doe.gov
	Green-e Renewable Energy Certification	www.green-e.org
	Energy Savers	www.energysavers.gov
	The Power is in Your Hands	www.powerisinyourhands.org
Procurement	Green Seal	www.greenseal.org
	Suppliers of Recycled Content Products	www.epa.gov/epaoswer/non-hw/reduce/wstewise
Action & education groups	Marine Stewardship Council	www.msc.org
	Certified Humane	www.certifiedhumane.com
	Idealist	www.idealist.org
	Social Marketing Institute	www.social-marketing.org/index.html
	Clean Air Council	www.cleanair.org
	Forest Stewardship Council	www.fsc.org
	"Green" Hotel Association	www.greenhotels.com
	Pay It Green	www.payitgreen.org
	National Resources Defense Council	www.nrdc.org
	Solar Living Institute	www.solarliving.org
	American Solar Energy Society	www.ases.org
	Carbonfund	www.carbonfund.org
	My Climate	www.myclimate.org
	Conservatree	www.conservatree.org
	Green Globe	www.greenglobe21.com
	Enivronmental Defense Fund	www.edf.org

websites

GreenRoofs.com	www.greenroofs.com
Green Source	www.greensource.construction.com
The World Bank	www.worldbank.org
Greenguard Environmental Institute	www.greenguard.org
Global Green USA	www.globalgreen.org

Recycling

Electronic Recyclers International	www.electronicrecyclers.com
Construction Materials Recycling Assocation	www.cdrecycling.org/
Recycling Batteries	www.rbrc.org
Plastics and the Environment	www.plasticsresource.com
NSF's Recycling Guide	www.nsf.org/consumer/recycling

Phone Recycling

Collective Good International	www.collectivegood.com
The Charitable Recycling Program	www.charitablerecycling.com
Wireless Recycling	www.wirelessrecycling.com
ReCellular	www.recellular.com
Wireless Foundation	www.wirelessfoundation.org

Computer Recycling

Earth 911	www.earth911.org
Electronics Industries Alliance	www.eiae.org

General green shopping

Eartheasy	www.eartheasy.com
Co-op America, National Green Pages	www.coopamerica.org/pubs/greenpages/
Ecomall—Environmental Shopping Center	www.ecomall.com
Global Exchange/Fair Trade	www.globalexchange.org
One Percent for the Planet	www.onepercentfortheplanet.org
Responsible Shopper	www.responsibleshopper.org
Earth Animal	www.earthanimal.com

Government agencies

Energy Star, U.S. Department of Energy	www.energystar.gov
Standby Power	www.standby.lbl.gov
The White House Council on Environmental Quality	www.whitehouse.gov/ceq/
U.S. Department of Energy	www.eere.energy.gov
U.S. Environmental Protection Agency, WaterSense	www.epa.gov/watersense
Materials Exchange Resources	www.epa.gov/jtr/comm/exchange.htm

Directories

EnviroLink Network	www.envirolink.org
Green Pages Co-op	www.greenpages.org
National Environmental Directory	www.environmentaldirectory.net

Investment	Social Investment Forum	www.socialinvest.org
	Progressive Asset Management	www.progressive-asset.com
	First Affirmative Financial Network	www.firstaffirmative.com
	Grameen	www.grameen-info.org
	Social Funds/SRI World Group, Inc.	www.socialfunds.com
Green Media	E/The Environmental Magazine	www.emagazine.com
	Earth Policy Institute	www.earth-policy.org
	Environmental Issues Newletter	www.environment.about.com
	Environmental Health News	www.environmentalhealthnews.org
	Environmental New Network	www.enn.com
	Grist Magazine	www.grist.org
	Tree Hugger, Online Magazine	www.treehugger.com
	The Green Guide	www.thegreenguide.com
	Environmental Design and Construction	www.edcmag.com
	EcoIQ Magazine	www.ecoiq.com
Transport	Electric Vehicle Assoc. of America	www.cvaa.org
	Enviornmental Guide to Cars and Trucks	www.greenercars.com
	Transportation Almanac—Energy, Pollution	www.bicycleuniverse.info
	Car Information—Mileage, Hybrids	www.fueleconomy.gov
	Center for Climate Change & Environmental Forecasting	www.climate.dot.gov
Case Studies	United States Green Building Council	www.usgbc.org
	IF Green	http://ifgreen.com/
	Whole Foods Market	www.wholefoodsmarket.com
	William McDonough + Partners	www.mcdonoughpartners.com
	F.W. Honerkamp Co., Inc.	www.honerkamp.com
	Design Solutions	www.sfdesignsolutions.com
	Taggart Construction	www.tagcon.com/
	Tepeyac Haven, Beacon Development Group	www.beacondevgroup.com/projects/ tepeyac.html
	Catholic Charities of Spokane	www.catholiccharitiesspokane.org/
	CU Green	http://green.mae.cornell.edu/

glossary

biodegradable> capable of decaying as a result of the action of microorganisms that break the material down into naturally recyclable elements.

biodiversity> all life on Earth, including the variability within it and between ecological communities or systems.

carbon emission> carbon substances like carbon monoxide and carbon dioxide that pollute the atmosphere and contribute to global warming.

carbon footprint> the impact a person or business has on the environment in terms of the amount of greenhouse gases produced, measured in units of carbon dioxide.

carbon sequestration> also known as carbon pooling or carbon sinking, carbon sequestration means capturing and storing carbon in forests, soils or in the oceans, so as to reduce the buildup of carbon dioxide in the atmosphere. Carbon sequestration is encouraged through enriching soils, oceans or underground geological repositories with elements that increase the uptake of carbon. Currently, increasing carbon storage in above-ground ecosystems is the most widely used, since it is the easiest and most immediate option.

carcinogen> a cancer-causing substance.

climate change> the variation in Earth's climate over time, largely involving temperature changes in the atmosphere. Scientists believe dangerous climate change is being caused by global warming, which is in turn significantly spurred on by greenhouse gas emissions.

compact fluorescent lights (CFLs)> a style of lightbulb that significantly reduces energy usage. Derived from the fluorescent tubes invented in the 1970s, CFLs emit the same amount of visible light, use less power and have a longer rated life. Some concern now exists around their safe disposal.

companion planting> the planting of different crops close to each other, in order to take advantage of the plants' natural properties to help the others.

conservation> sustainable use and protection of natural resources including plants, animals, mineral deposits, soils, clean water, clean air, and fossil fuels such as coal, petroleum and natural gas.

dioxin> the popular name for a family of organic compounds that bio-accumulate with toxic effect in humans and wildlife. Two of the most widely studied sources of dioxins are the making of the herbicide Agent Orange and the chlorine bleaching of wood pulp in paper-making.

double glazing> window treatment in which air is trapped between two panes of glass, creating an insulating barrier that reduces heat loss, noise and condensation.

ecological footprint> the amount of biologically productive land and sea area needed to generate the resources a human population consumes, and to absorb the corresponding waste.

ecosystem> the interaction of plants, animals and the environment.

embodied energy> the production of a building or product requires a certain amount of energy, ranging from the mining and manufacturing of raw materials to transporting them and even the administrative processes involved in producing it; also, maintenance or renovation can add further embodied energy. An assessment for embodied energy takes into consideration all those processes.

Energy Star rating> a mandatory government labeling system for some (not all) electronic appliances: the higher the number of stars, the more efficient the appliance. Products labeled with an Energy Star meet strict energy efficiency guidelines set by the EPA and US Department of Energy.

e-waste> discarded electrical equipment such as cell phones, computers, DVD players and cabling.

Fairtrade> a certification system that labels products that meet "fair" environmental, ethical labor and developmental standards.

food miles> the distance food travels from the source of its growth to the consumer's table.

formaldehyde> a chemical often used in disinfectant because of its bacteria-killing properties, but which has also been classified as a carcinogen.

genetic engineering> directly manipulating the genes (DNA) of an organism in order to change its character. Concerns about genetic engineering include disruption of natural ecosystems and the unknown long-term effects of genetically modified crop consumption.

global warming> the accelerated warming of Earth's surface due to release of greenhouse gases generated from industrial activity and deforestation. The possible results of global warming include rising sea levels, increasing intensity of extreme weather events, changes in amount and pattern of rain and snow, changes in crop yields and ocean trade routes, glacier retreat, species extinctions and increases in disease.

green electricity> electricity obtained from renewable sources, such as wind, sun and water, not from fossil fuels such as coal or oil.

greenhouse effect> Earth's atmosphere allows solar radiation to be absorbed by the planet's surface, which is then re-emitted as heat. This heat is in turn reflected back by gases such as carbon dioxide, methane, nitrous oxide and ozone that are in the atmosphere (greenhouse gases). This is the greenhouse effect that keeps Earth warm. The increasing release of greenhouse gases into the atmosphere is causing global warming.

greenhouse gases> gases such as methane and carbon dioxide (CO_2) that contribute to the greenhouse effect.

gray water> gray water is fresh drinking water that has been used for cleaning in the kitchen, laundry or bathroom (not the toilet). It comprises 50–80 percent of residential waste water.

heat gain> the heat accumulated in a building through different sources such as outdoor temperature and humidity levels, and also people inside it, lights, computer, copiers, fridges and ovens. Most of the heat is usually gained through the sun beating down on the roof and pouring through windows.

indoor air quality> the purity of the air in an enclosed space. There are many sources of pollution in any indoor space, much of it deriving from building materials and furniture treatments, but also from insulation, cleaning products, pesticides, paints and even outdoor pollution. Inadequate ventilation, heat and humidity levels can also contribute to indoor pollution.

insulation> insulation consists of material applied to ceilings, walls, floors and the roof to reduce the rate of heat transfer through the external surfaces of a building—like wrapping a blanket around a house. Insulation can consist of natural or synthetic materials, or a combination of both.

glossary

Experts distinguish between two types of insulation: bulk and reflective. Bulk insulation, as the name implies, adds extra mass to the walls to stop heat transfer, while the second type of insulation literally reflects the heat.

kWh> a kilowatt hour is a unit of energy measuring 1000 watt hours; the amount of energy produced or transferred in one hour by one kilowatt of power.

landfill> a site where waste is placed in the ground. This can happen in a controlled or uncontrolled way, and usually refers to waste that cannot be recycled but is left to decompose.

LED> this stands for light emitting diode. LEDs are semi-conductor diodes that typically emit a single wavelength of light when charged with electricity. First introduced in the 1970s, LEDs use less energy than simple incandescent light globes, are virtually indestructible and last for decades.

life cycle analysis (LCA)> all building products can affect the environment in many ways during every stage of their life, whether this is in production or after it's been discarded. A life cycle analysis looks at all these stages "from cradle to grave,"

beginning with the acquisition of raw materials, manufacturing and shipping processes, and its effects throughout its use (such as indoor air quality, durability and performance), as well as how easy it is to recycle or reuse the material at the end of its life.

methane> a gas emitted by organic matter breaking down.

old-growth forest> an ancient and unique area of forest that cannot be replicated and has not been destroyed by logging. It often contains rare fauna and flora species, and stores large amounts of greenhouse gases.

operational energy> the energy that is used to operate a particular device or accumulation of devices. The energy used by all the appliances, light sources and heating in a home amounts to the operational energy of the home.

organic> produced without fossil fuel–based fertilizers, synthetic pesticides or genetically modified crop varieties.

orientation> refers to the way a new building is positioned on a site in relation to cardinal points and the sun. The first principle for building a sustainable home is ideal orientation

so as to maximize the benefits derived from sunlight and passive heating, as well as prevalent wind directions and shade. Good orientation makes a building more comfortable to live in and cheaper to run.

ozone layer> the layer in Earth's atmosphere that absorbs harmful ultraviolet radiation from the sun. A depleting ozone layer allows more radiation to pass through, causing genetic damage to life on Earth. Ozone levels are depleted by the release of gases like chlorofluorocarbons (CFCs) contained in products such as CFC aerosol sprays, now banned in most countries.

passive design> passive design is design that does not require mechanical heating or cooling, instead drawing advantage from a home's orientation and spatial configuration. Homes that are passively designed take advantage of natural energy flows to maintain thermal comfort.

passive solar technologies> technologies that convert sunlight into usable heat for water or air, cause air movement for ventilating, or store heat for future use without the assistance of other energy sources. They do so without active mechanical systems, as opposed to active solar systems, which use pumps or fans to increase usable heat in a system.

permaculture> a garden or farming system that uses a variety of plants to encourage self-sufficient crop production.

photovoltaic power> photovoltaic panels, installed on roofs or near homes, generate electricity using the sun's light, unlike the solar panels used for heating water, which absorb heat. A panel consists of multiple cells made up of semi-conducting silicon and when this is exposed to light, electrical charges are generated that can be conducted away by metal contacts as direct current.

product miles> the amount of energy needed to get a product to the customer. The term is used as a measurement of the environmental impact of bringing any product to market.

renewable resource> a renewable resource is one that can be replaced over time by natural processes, such as fish populations or natural vegetation, or is inexhaustible, such as solar energy.

self-sufficient> able to provide for one's own needs without external intervention or assistance.

sick building syndrome> a term used to describe physical discomfort felt by building occupants, often with severe implications on their health, that can be attributed to the building, its structure or furnishings. Often, sick building syndrome is related to indoor air quality.

solar power> system of collecting solar energy (from the sun) to generate electricity; power created by converting sunlight into electricity.

sustainable> a sustainable process or state is one that can be maintained in its original form indefinitely. That is, a system that will not collapse or break down. For example, creating energy from fossil fuels that will eventually run out is not a sustainable practice.

sustainable product development (SPD)> a method of product development that improves efficiency and lowers environmental impact by reducing the impact of raw materials, processing and waste.

thermal mass> generally, any material that has the capacity to

store heat. When applied correctly, thermal mass can significantly reduce the requirement for active heating and cooling systems and the consumption of active solar, renewable energy and especially fossil fuel technologies.

volatile organic compounds (VOCs)> these are the fumes that evaporate from conventional pesticides, cleaning products, paints, finishes and glues. Harmful to animal and plant life, they can also jeopardize human health and make a significant contribution to indoor air quality. VOCs are held responsible for causing sick building syndrome.

WaterSense Program (Environmental Protection Agency)> the U.S. government system for labeling the efficiency of water-using products such as showerheads. A WaterSense label means that whatever product or program it appears on meets water efficiency and performance criteria.

Clean up the world

about Clean Up the World

Clean Up the World, the international outreach campaign of Clean Up Australia, was co-founded by *True Green* creator Kim McKay and Ian Kiernan, AO—legendary yachtsman and 1994 Australian of the Year.

In partnership with the United Nations Environment Programme (UNEP), Clean Up the World annually attracts more than 35 million volunteers who join community-led initiatives to clean up, fix up, and conserve their local environment.

Fifteen years after its launch, the campaign has become a successful action program that spans more than 120 countries, encouraging communities to take control of their own destiny by improving the health of their community and environment.

Global activities include waste collection, education campaigns, environmental concerts, creative competitions, and exhibitions on improving water quality, planting trees, minimizing waste, reducing green house gas emissions, and establishing recycling centers.

Participants range from whole countries (e.g. Australia and Poland), community and environmental groups, schools, government departments, businesses, consumer and industry organizations, to sponsors and dedicated individuals who either work independently in their local communities or with other groups in a coordinated effort at a regional or national level.

Visit the Clean Up the World website to find out how your community, company, or organization can become involved: **www.cleanuptheworld.org.**

"We can all make a difference and joining in Clean Up the World is a simple and practical way to do something about global warming and climate change. Be part of the solution and help Clean Up Our Climate!"

Ian Kiernan, AO
Chairman & Founder, Clean Up the World

Kim McKay, AO (right) is co-founder and deputy chairwoman of Clean Up Australia and Clean Up the World. An international social and sustainability marketing consultant, Kim counts National Geographic among her many clients. In 2008 she was made an Officer of the Order of Australia for distinguished service to the environment and the community.

Jenny Bonnin (left) is a director of Clean Up Australia and Clean Up the World. She and Kim founded Momentum2, a social and sustainability marketing firm, and together created the True Green brand. Jenny has recently embarked on a role with the Clinton Climate Initiative in Sydney to further her passion for creating a sustainable future. She has two children and two stepchildren.

Kim and Jenny's previous books—*True Green: 100 Everyday Ways You Can Contribute to a Healthier Planet* (ABC Books, 2006); *True Green @ Work: 100 Ways You Can Make the Environment Your Business* with business writer Tim Wallace (ABC Books, 2007); and *True Green Kids: 100 Things You Can Do to Save the Planet* (ABC Books, 2008)—have also been published in the United States by National Geographic Books.

Marian Kyte is a freelance designer and creative director of True Green. She has a passion for incorporating sustainability principles into her work. Her clients have included Qantas, Craftsman House Books, Power Publications, Sherman Galleries, *Art & Australia*, *Limelight* magazine and True Green. Her son Locky is her inspiration.

Vivanne Stappmans is a researcher and writer with degrees in journalism and design, who has spent years writing about places, people and homes. Well versed and ever curious about design, Vivanne has explored, interviewed, written and edited for many of Australia's leading publications.

acknowledgments

True Green Home has helped consolidate our True Green family of colleagues and supporters. We could not hope to do a True Green book without Marian Kyte, who is an endless source of creative inspiration and joy. It never ceases to amaze us how she can illustrate our points in the most colorful, simple and brilliant ways.

Vivanne Stappmans managed to deliver the research and a beautiful baby girl, Ivy, at the same time! She's an incredibly clever and generous writer who cares deeply about her daughter's future. Huge thanks also to Nina Fedrizzi, who reversioned the book for U.S. readers and brought a great knowledge base and creative insight to the project. Nina Hoffman and Kevin Mulroy of National Geographic Books are friends and champions who have believed in True Green from the outset. Special thanks to our dedicated and patient editor, Olivia Garnett, along with the entire NG Books publicity team.

We especially want to thank our erstwhile marketing coordinator, Kylie Guthrie, who has kept her eye on the ball though the book's planning and production and who is a rock in our office. We also thank the experts who contributed their insights for the feature pages of the book and salute the green architects and designers of America for helping to transform the way we live. Thanks again to our fellow Clean Up directors and the Clean Up staff for their continued support.

We dedicate this book to the favorite builder in our lives, Ian Kiernan, who loves nothing more than brandishing his tape measure on a building site and creating beautiful restored buildings from virtual rubble. He is a true builder in so many ways, and his dedication to building a better future for the global environment through Clean Up and countless green technology projects is saluted.

"Our personal consumer choices have ecological, social, and spiritual consequences. It is time to re-examine some of our deeply held notions that underlie our lifestyles."

David Suzuki, award-winning scientist, environmentalist and broadcaster

Elysium House 7 by Andrew Maynard Architects. Rendering: Virtuocity